Orchids *of* Asia

Teoh Eng Soon

Times Editions
Marshall Cavendish

Front cover: *Grammatophyllum multiflorum* var. *tigrinum*
Back cover: *Dendrobium thyrsiflorum*
Title page: *Dendrobium simillieae*
Page 4 (facing Contents): *Dendrobium* Anucha Flare x Chaisri Gold
Page 5 (Contents): *Dendrobium anosmum* var. *alba*

Designer: Benson Tan
Illustrator: Lim Thiong Ghee
All photographs were taken by the author, unless otherwise stated.

© **2005 Teoh Eng Soon**

First published in 1980 as *Asian Orchids*
Second edition published in 1989 by Times Books International

This edition published in 2005 by Times Editions – Marshall Cavendish
An imprint of Marshall Cavendish International (Asia) Private Limited
A member of the Times Publishing Limited
Times Centre, 1 New Industrial Road, Singapore 536196
Tel: (65) 6213 9300   Fax: (65) 6285 4871
E-mail: te@sg.marshallcavendish.com
Online Bookstore: www.marshallcavendish.com/genref

Malaysian Office:
Marshall Cavendish (Malaysia) Sdn Bhd (3024-D)
(General & Reference Publishing)
(Formerly known as Federal Publications Sdn Berhad)
Times Subang
Lot 46, Persiaran Teknologi Subang
Subang Hi-Tech Industrial Park
Batu Tiga, 40000 Shah Alam
Selangor Darul Ehsan, Malaysia
Tel: (603) 5635 2191, 5628 6888 Fax: (603) 5635 2706
E-mail: cchong@my.marshallcavendish.com

Limits of Liability/Disclaimer of Warranty: The author and publisher of this book have used their best efforts in preparing this book. The publisher makes no representation or warranties with respect to the contents of this book, and specifically disclaims any implied warranties or merchantability or fitness for any particular purpose, and shall in no events be liable for any loss of profit or any other commercial damage, including but not limited to special, incidental, consequential, or other damages.

National Library Board Singapore Cataloguing in Publication Data

Teoh, Eng-Soon, 1938-
    Orchids of Asia / Teoh Eng Soon. – 3rd ed. – Singapore : Times Editions-Marshall Cavendish, 2005.
    p. cm.
    First ed. published as: Asian orchids. Times Books International, 1980.
    Includes bibliographical references and index.
    ISBN : 981-261-015-4

1. Orchids – Asia. 2. Orchids – Varieties – Asia. 3. Orchid culture.
I. Title.

SB409.5
635.9344095 -- dc21   SLS2005005377

Printed in Singapore by Saik Wah Press

For Phaik Khuan,
John, Kristine,
Chrissy and Ning

CONTENTS

Acknowledgements

I value the opportunity to judge at the shows organised by the Orchid Society of Southeast Asia, which have helped me keep abreast of the happenings in the orchid world and stay in touch with my friends during that period when the demands of my practice, my professional interests, and my dogs caused a decline in my interest in orchids. My wife and I delight in taking strolls through the National Orchid Garden, its Mist House and Cool House with their varied displays of orchid species and hybrids. We treasure the publications that have taught us much of what we know of orchids, in particular the *American Orchid Society Bulletin* and the *Malayan Orchid Review*, to which we have subscribed for over 35 years. My thanks to the people who have thus enriched our lives and fostered our interest in orchids.

Our orchid friends from many parts of the world have been generous to us in many ways, and they are remembered with much fondness even though they cannot all be mentioned by name. I am particularly grateful to the following people who have assisted me during the preparation of this revised edition: Mak Chin On, Nopporn Buranaraktham, Veradej Boonyuenvetwat, Dr. Kiat Tan, Dr. Tim Yam, Syed Yusof Alsagoff, John Elliot, Teo Peng Seng and Puah Gik Song. For the use of pictures, I thank Professor H. Kamemoto, Professor Rapee Sagarik, George Alphonso, John and Kristine Teoh, and the Orchid Society of Southeast Asia. I am grateful to Ms. Magdalene Chng who retyped the text of the second edition on my computer.

It has been a great pleasure to work with the staff of Marshall Cavendish on this book, and I would like to thank, in particular, my editor, Leong Wen Shan, and designer Benson Tan.

To a large extent, this book owes its existence to the devotion and tolerance of my wife, Phaik Khuan, and the support of my son, John.

Teoh Eng Soon
March 2005

Opposite: *Dendrobium* Anucha Flare x Chaisri Gold (mutation)

Right: A new phalaenopsis-*Dendrobium* mutation

Preface to Third Edition

Orchid growing in Southeast Asia, indeed throughout the world, has seen many changes over the past two decades. Logging and the conversion of virgin jungle into dams and human settlements in Southeast Asia yielded many new species of orchids which have been described, collected and distributed. New and spectacular species of orchids, in particular *Paphiopedilum* and *Renanthera*, were discovered when studies were undertaken on the fauna and the flora of the horticulturally famous areas of South China and Indochina.

A new generation has taken over from the old; there are new hobbyists, new collectors of fine plants and many new nurserymen. Taiwan and Japan are major new players. Within a few years, the People's Republic of China is likely to become an inexhaustible source of low-priced orchid hybrids. Large agricultural corporations have injected tremendous expertise, finance and personnel into the making and propagating of new hybrids, thus accelerating the process. Nevertheless, they have left many areas untouched, and the skilful hobbyist breeder always managed to find a niche where he could produce good work. A quality orchid can now be purchased for a song, thanks to clonal propagation undertaken in Asia. The process further produced an array of novel mutant orchids with interesting patterns in several genera of popular orchids. Some of these mutant characteristics are transmissible, thus the potential for improvement appears limitless.

The last edition of *Orchids of Asia* was published in 1989 and has been out of print for many years. It received many enthusiastic reviews. One writer said that she kept the book at her bedside so that she could instantly refer to it whenever an idea came to her. Judges told me that it was a useful reference and guide. Several colleagues found the tone refreshing: they were comfortable with the style. Many liked the photographs. With these comments in mind, I decided that this new edition should present a historical perspective on the development of orchids in Southeast Asia, showcasing the progress in hybridising over the past 75 years, as well as explaining the basis for growing orchids the way we do. Changes have been made to the text, where necessary, to update the information. New species and new hybrids that impressed the author are also introduced. In addition, the work on *Cattleya*, *Dendrobium* and *Vanda* in Thailand, *Doritaenopsis* and *Phalaenopsis* in Taiwan, and *Aranda* and *Mokara* in Singapore and Malaysia are highlighted.

A new chapter describes the amazing somatic mutations that have appeared in *Cattleya*, *Phalaenopsis* and *Dendrobium*. It replaces the chapter "Orchids as Aphrodisiac, Medicine and Food", which was a bit whimsical. Agreeing with the adage that a picture speaks a thousand words, more than 300 additional photographs have been included in this new edition of *Orchids of Asia*.

The author hopes that this book succeeds in simplifying the concepts of orchid growing and that it will induce its readers to bring orchids into their lives.

Teoh Eng Soon
March 2005
Singapore

Opposite: *Eria rhynchostyloides*

Preface to Second Edition

When *Asian Orchids* was released in Australia, a television book reviewer remarked that there was enough material in it to fill three books. I have not followed his suggestion to split the work into several parts. Instead, for this second edition, three new chapters have been added to cover commercial orchid farming, the handling of cut flowers and the quaint use of orchids as medicine or food. The other chapters have been revised to bring them up to date. Half of the photographs have been replaced by new ones.

I have decided to retain the original character of *Asian Orchids* because this is what makes this book unique. Its emphasis is on warm-growing orchids. Science is interpreted in lay terms and combined with practical growing experience. The amateur's enthusiasm is tempered by close association and long discussions with professional growers. The magic of the photographic lens should continue to enthrall the orchid buff.

I wish to record my gratitude to Professor Eric Holttum, who sent me a list of corrections after reading through my first edition; his scholarship and charity have been a source of inspiration.

Finally, I hope this second edition will give as much pleasure to its readers as the first one did.

Teoh Eng Soon
January 1989
Singapore

From Left: *Aranda* Majula; *Vanda* Thanachai x *Rntda.* Nancy Chandler; *Phalaenopsis* John Teoh; *Stamariaara* Noel

Opposite: *Phalaenopsis cornu-cervi alba*

Preface to First Edition

In the continent of Asia, no country has played a more significant role than Singapore in the promotion of orchid cultivation. Its first orchid hybrid of 1893, the *Vanda* Miss Joaquim, pioneered the international cut orchid industry and provided impetus to commercial orchid growing in Hawaii. Over the past 50 years, the Singapore Botanic Gardens, working initially with the Malayan Orchid Society (founded in 1928 and now renamed the Orchid Society of Southeast Asia), has created numerous tough but exquisite orchid hybrids which travel daily to Europe, the United States, Japan, Hong Kong and Australia to grace homes, offices and hotels. The Singaporean orchidists worked on the vast pool of orchid species from the entire Southeast Asian region, including Papua New Guinea. They have extended the early hybridisation work in Java and have made good use of the more recent contributions, both in plants and technology, from Hawaii, continental America, Thailand and Malaysia. The ability of the modern Asian tropical hybrids to bloom continuously and the ease of their cultivation have turned the exotic orchid into a favourite garden plant.

Yet, the only orchid books to emerge from Singapore have been the outstanding botanical treatise, Holttum's *Flora of Malaya: Orchids*, which was published in 1953, and the two volumes giving descriptions of Malayan orchid hybrids which appeared in 1956 and 1963. Orchid culture has taken great strides during the past 15 years. Today, we can also take advantage of advanced colour printing technology to display the beauty of the orchid bloom side by side with the technical data.

In this book, I have attempted to combine a quick review of the more horticulturally important orchid tribes of the Asian lowlands with basic essential information on orchid needs, hybridisation, diseases, a simplified approach to orchid culture and the aesthetic appreciation of the orchid. It is not a botanical work. It is written as an introduction to the wonderful world of Asian orchids for the grower and for people who love flowers.

I have not included a comprehensive review of the work on orchids which has been done in Singapore and other parts of Asia. Thus, there is no extensive listing of Asian hybrids. Instead, the chapter on hybrids emphasises the scientific elements of hybridisation because the knowledge of the genetics involved is far more than the names of the parent plants and their progeny. I apologise to the beginners, who may find some difficulty getting used to the technical terms in certain chapters. But as beginners often become thoroughly immersed in their hobby, they may glean some useful data from these very sections at a later re-examination.

The older growers tell me that they miss the species and the early hybrids which are no longer included in modern amateur collections. They are still grown at the Singapore Botanic Gardens, unassuming plants which make delightful photographs. I have tried to include some examples of such orchids.

The best pictures of flowers take on an aspect of 'glamour photography' in always emphasising the loveliness of the subject instead of being merely accurate reproductions. This was the approach I strove to follow when I started to do photographs for this book. The photographs were taken on location at the Singapore Botanic Gardens, at various commercial nurseries in Singapore, in the home of friends and at the author's home. There was no opportunity to use studio flood-lighting and sometimes it was not possible to isolate the plant for photography. In selecting the orchid pictures for this book, I have based my choice on their pictorial value rather than on their merit based on the strict criteria for the award judging of orchids, but this is not being done to express any disagreement with the award criteria or to show my personal preference for the orchids portrayed. Many beautiful orchids do not appear simply because I did not have the opportunity to photograph them or my slides have not turned out well. Because of the constraints of time, it has not been possible to photograph every orchid which I would like to show or to wait until each plant reaches its prime before its flowers are photographed. I have included a few slides from my friends, Professor H. Kamemoto, David Lim and George Alphonso, when they helped to amplify the text.

Enjoy the orchids of Asia.

Teoh Eng Soon
June 1980
Singapore

Left: For over 15 years, the author has been growing this pink *Phalaenopsis* and other orchids on a fibreglass rockery which borders the driveway at his home. The *Phalaenopsis* is shaded from direct overhead sun by a bird's nest fern (*Asplenum nidus*) but it receives bright sunlight reflected from a white wall on the opposite side of the driveway. The *Phalaenopsis* has become very hardy and stays in bloom for many months.

Opposite: *Phalaenopsis amabilis* is also grown on a fibreglass rock in the Cool House of the Singapore Botanic Gardens. Although the plant is not very large, the cool temperature, strong light and misting have caused it to produce a magnificent spray of flowers.

Chapter 1

Introduction

The large orchid family, comprising some 25,000 species, is one of the most widespread of plant families. Orchids grow in dark tropical jungle, on the exposed slopes of rock cliffs, clinging to rocks on the shoreline just above high tide, at the edge of the desert, on the foothills of the Himalayas, and even in the Arctic. In 1976, some members of the Orchid Society of Southeast Asia based in Singapore found an orchid plant bursting through the macadamised driveway of the hospital where their meetings were being held.

When orchids were first introduced into England in the 19th century, the editor of the *Botanical Register* remarked: "The cultivation of tropical parasites was long regarded as hopeless; it thus appeared a vain attempt to find substitutes for the various trees each species might effect within the limits of a hothouse." He had the mistaken notion that orchids were parasites similar to the mistletoe.

Orchids, of course, are not parasites. They are mostly epiphytes — plants which depend on trees for support but not for nutrition and the rest are terrestrial plants, some growing on leaf mould. Chinese gardeners have cultivated *Cymbidium* in pots with soil for 5,000 years, but they grow *Vanda* and *Aerides* suspended in baskets. Alvim Semedo, a Jesuit missionary, saw these hanging orchids during his travels in China in 1613. He referred to them as *diao hua* 'hanging flowers', or aerial plants that possessed the peculiar property of growing suspended in air. The Japanese nobility grew *Neofinetia falcata* in a similar fashion to enjoy their fragrance. In their native Thailand, *Vanda* and *Aerides* are still grown in empty hanging baskets, with their roots trailing a metre or two

in the air. Indeed, this is one of the better ways to grow such orchids.

Orchids come in many shapes and colours. The resemblance of orchids to moths (*Phalaenopsis*), butterflies (*Oncidium papilio*), spiders (*Arachnis*), bees (*Ophrys*), slippers or moccasins (*Paphiopedilum* and *Cypripedium*), or dancing ladies (*Oncidium*) led to their numerous fanciful names. While most cultivated orchids are flamboyant and dazzling, the vast majority of native species are modest, and many only reveal their beauty when they are examined under a magnifying lens.

There are several orchid species which are completely leafless, depending on their green roots to produce their food while their stems are merely stumps to hold the flowers. A handful of species are subterranean and live as saprophytes, relying on decaying organic matter for food; only their flower stems emerge from the ground to permit pollination and ensure the survival of their species.

Indeed, the variation in form among orchids is so diverse that at first glance one may not realise that all the individual species of orchids belong to the same family, the common denominator being the essential reproductive part of the flower.

The different orchid species classified within a genus (and related genera) can usually be cross-bred, and this sometimes occurs in nature with the help of non-discriminatory insect pollinators. Cross-pollination by man is far more effective and almost 100,000 hybrids have been produced in the last century.

Orchids mean different things to various people. They have been the hobby of emperors and kings, special flowers for the wedding bouquet or for Mother's Day, herbs from which to obtain a perfume or flavouring for chocolate and ice cream, a lotion to bestow strength, even an aphrodisiac.

During the Middle Ages, it was a common belief in Europe that orchid plants, or *Satyria*, grew in places where cattle, sheep or horses mated. According to a legend retold by George Ure Skinner, the very first orchid was the son of a nymph and a satyr, a fellow of unbridled passion. At a festival of Bacchus, intoxicated with drink, Orchid attacked a priestess whereupon the entire gathering fell upon him and tore him limb from limb. As a favour to his pleading father, the gods changed him into a flower which retained his nature. Thus, it was believed that eating the plant's roots aroused a satyric passion.

The word orchid is derived from the Greek *orchis*, meaning testis, an association made by Theophrasus (370–285 B.C.), whose attention was drawn to the similarity between the underground bulbs of the Mediterranean *Orchis* and the mammalian organ. Carl von Linnaeus, the founder of modern plant taxonomy, retained this name in his *Species Plantarum*, the origin of botanical binomial classification.

To the Chinese, the orchid, or *lan hua* (*Cymbidium ensifolium*), is the symbol of the scholar: unassuming, enduring, chaste and ascetic. ("Tall grass may hide a fragrant orchid: a thatched cottage may cover the heir to a throne.") Orchids were also symbols of love and beauty and stood for grace, refinement, fragrance and all things considered noble and elegant in a woman. It is a favourite plant for expressing the rhapsodic strokes of the Chinese brush: "Plant bamboo when you are angry; orchids when you are happy."

Confucius (551–479 B.C.) said "acquaintance with good men was like entering a room full of fragrant orchids." During the Yuan Dynasty (1279–1368), one of the darkest periods of Chinese history, Cheng Ssu-hsiao (circa 1250–1300) lamented the rape of China by the Mongols in an exquisite painting of uprooted orchids, surely one of the most oblique expressions of political protest ever recorded. Later, when forced to paint in prison,

he retorted, "You may have my head, but not my orchids!"

Orchid cultivation was a popular pastime of scholars and the gentry in the preceding Sung Dynasty (960–1279). Chao Shih-keng published *A Treatise on Orchids of Chin-Cheng* in 1233, describing culture methods for 22 kinds of white- and purple-flowered orchids which included *Cymbidium*, *Phaius* and *Calanthe*. In the southern Fukien province, Wang Kuei-hsueh produced another comprehensive work on 37 species entitled *Treatise on Orchids*, published in 1247. *Changes of Orchids* by Ting Wong appeared in 1250.

The promise of beautiful orchids has lured men and women to brave the perils of virgin jungle, to scale mountains and wander into ravines in the quest for these exotic flowers. Neither the presence of pirates and cannibals, the story of man-eating orchids, nor earthquakes and hurricanes deterred the orchid hunters of the 19th century. When these jungle plants reached London, the auction houses of Protheroe and Morris were stormed by hundreds of top-hatted orchidophiles arriving post-haste in their cabs to bid for the mountains of orchid plants put up for sale. In the 1950s, the appearance of the fabulous *Vanda* Tan Chay Yan in Singapore revitalised orchid interest throughout Southeast Asia. Wealthy amateur growers vied with one another to own and

Opposite: A *Cattleya* hybrid bred in Thailand and intended for export to the United States and Europe. However, it can thrive in the tropical lowlands and flower freely — about two to three times a year. *Cattleya* can be grown in the open in Singapore and it does not demand much care, making it an ideal garden plant. There has been much success in the hybridisation of *Cattleya* and its allies in Taiwan and Thailand, although these are not native Asian genera.

flower the newest hybrids, and many rushed from air-conditioned offices to humble orchid nurseries in pursuit of the latest and most expensive orchids from France and Hawaii.

The mad scramble for new species or hybrids is over. Mericloning, the first of genetic engineering's commercial successes, has made available vast quantities of top-quality orchids, causing the price of prized orchid plants to plunge. An orchid seedling today may cost less than a cup of coffee in a hotel and a robust, flowering plant sells for the price of a music CD. In Singapore, a lovely spray of orchids costs 50 cents (US 30 cents). Orchids are no longer the playthings of the wealthy. In terms of money, everyone can afford to grow orchids. Nevertheless, some orchids do carry fancy price tags; and others may be fastidious, requiring hothouses or cool houses, depending on the location of the grower. It is not necessary for the novice to start with such plants. With some basic knowledge and a little care, it is easy to find the correct orchids to start the hobby of orchid cultivation.

My own interest in *Phalaenopsis* was sparked off by the impact of a fabulous orchid show which my wife and I visited at the Bayfront Auditorium in Miami. Although our permanent home was halfway across the world in Singapore and we were scheduled to stay on for only another year in Florida, the attraction of the orchids was so great that we decided to purchase a community pot of *Phalaenopsis* and a flask of *Oncidium* on the spot. Soon we were spending weekends visiting nurseries and private growers in Miami, then further afield in Florida, and adding to our collection of flasks and community pots. Being then a young academic, and in order to conserve our funds, I split some of the community pots and flasks with a friend. We had no garden; all our seedlings were grown on the window

Opposite: *Dendrobium macrophyllum* flowering at the Cool House of the Singapore Botanic Gardens

Right: *Cymbidium finlaysonianum*, grown in the ancient manner — in a coconut husk and hanging from a tree

sill, in the kitchen, the lounge and, finally, the bedroom. Despite the air-conditioning, the seedlings thrived, never showing any sign of wilting. I had planned a long lecture tour and wanted to visit the top medical centres in the United States on the homeward journey to Singapore. In addition to two small children and luggage, we travelled with a dozen flasks of orchid seedlings which we carefully unpacked and placed by the window whenever we checked into a hotel. We left our community pots and larger seedlings in the care of the late Roy K. Fields, who shipped the plants to us when we arrived in Singapore. We did not manage to establish a proper garden for three months after our return, and the orchids were grown in a spare bedroom during this period. Nevertheless, our first orchid bloomed the following Christmas and our *Phalaenopsis* won award after award in Singapore.

Our personal, initial experience with these delightful plants taught us several basic truths regarding orchids:

1. For many who venture into the world of orchids, the lure of new hybrids is irresistible because of the joy and fascination in watching for the first blooming of a new hybrid.
2. Orchid plants are easy to grow provided one understands their requirements.
3. Orchids can be adapted to many situations and can be grown very satisfactorily under varied conditions, as long as the plants are provided with sufficient light, humidity and a little air movement.
4. It does not cost a lot to own a fine collection of orchids.

The orchids which have been most widely written about are the *Cattleya*, *Odontoglossum* and *Miltonia*, which originate in the mountains of South America, and the *Cymbidium* and *Paphiopedilum* which come from the highlands of tropical Asia. These are also the genera which are most suited to the temperate climate. Such orchids have been bred to perfection in Europe, the United States and Australia. In the flower market today, one sees a different range of orchids — *Dendrobium*, *Vanda*, *Phalaenopsis*, *Aranthera* and the Scorpion Orchids, which have been bred in Singapore, Malaysia, Thailand, Indonesia, and now Taiwan. The accent in this book is on these tropical lowland orchids, the toughest and most beautiful members of the family.

Three examples of orchid types which are well suited for tropical lowland cultivation:

Opposite: *Aerides* Amy Ede

Below: *Vanda* Thanachai x *Renantanda* Nancy Chandler

Right: *Aranda* Baby Teoh

However, many orchids from other parts of the world were introduced into Southeast Asia more than 80 years ago and they are now a part of the garden flora. Breeders in Thailand have honed their hybridisation skill in *Cattleya* and other South American species. I have included some examples of their achievements.

With the construction of a Mist House and, more recently, a Cool House at the National Orchid Centre of the Singapore Botanic Gardens, it is now possible for the public to view many cool-growing orchids in spectacular floral display during their blooming season. The author has enjoyed his numerous visits to the Cool House and the pages of this revised edition bear testimony to those visits. It is difficult to collect good pictures of every representative species, but the recent displays at the Singapore Botanic Gardens and a general interest in orchid species among Singapore growers have certainly helped to enrich the collection in this volume. Perhaps a familiarity with the beautiful species of orchids will promote their conservation.

Orchids in their simulated 'natural' environment:

Left: The cool house simulates a cool mountain ambience.

Below: *Vanda* at home on a Frangipani tree

Chapter 2

The Orchid Plant and Its Flowers

Adaptation to extremely diverse environments has produced great variation in vegetative form among the 25,000 members of the orchid family. The uninitiated may find it difficult to recognise some of its member species as orchids when they are not in bloom. However, the flowers of nearly all orchids are constructed on a common plan, with certain distinctive features. The column and the lip (or labellum) are the hallmarks of the orchids.

On the evolutionary scale, orchids are very advanced plants. According to a theory proposed by Robert Brown in 1833 and supported by Charles Darwin in 1877, the orchids were probably derived from primitive plants whose flowers resembled the lily. The lily flower has 15 parts arranged in five whorls. Starting from the outermost layer, there is an outer ring of three sepals and an inner ring of three petals which alternate with the sepals. The sepals and petals are similar in shape and colour and they enclose the reproductive parts consisting of six stamens arranged in two whorls which in turn surround the three stigmata. The stigmata are attached by the style to a three-chamber ovary.

In the orchid flower, there are three sepals which are usually different from the petals in

shape but not in colour. There are also three petals but one has been greatly modified into a lip, an extremely complex structure which is quite different from the other two petals. The lip is often embellished with crests, horns, hair, ribs and other protuberances and is further accentuated by vivid coloration. Usually, the lip is the largest part of the flower but in some species the lip is small. The stamens, style and stigma are fused together into a single unique structure known as the column. In 99 percent of orchids, only the uppermost anther is retained and the pollen is clumped together into masses known as pollinia: these orchids belong to the subfamily *Monandrae* (meaning 'one anther'). About 1 percent of orchids have two functional stamens which are derived from the original two stamens situated on either side of the lip, both belonging to the inner whorl. These orchids are known as *Pleonandrae* (meaning 'several stamens'). The Slipper Orchids (*Paphiopedilum*) belong to this subfamily. The remaining three anthers which were present in the primitive 'lily' normally never appear in the orchid, having been lost in the evolutionary process; occasionally, they may reappear in freak flowers.

The flower of the *Cattleya* has narrow identical sepals. The top one is known as the dorsal sepal and the two spreading sideways on either side of the lip are the lateral sepals. In front of the sepals are the three broad petals, but one is so unique that it is distinguished by a separate name, that is, the lip (or labellum). The lip has three lobes, with the two sidelobes curling over in front to form a funnel which encloses the column. The midlobe of the lip is very colourful in the *Cattleya*, where it is

There are remarkable variations in the shape of orchid flowers, but they share a few things in common — a beautifully elaborate lip, the column and the pollinia.

Opposite: *Paphiopedilum bellatulum*, a Slipper Orchid

Right, Top: *Blc.* Memoria Helen Brown

Right, Bottom: A *Cleisostoma* species

Table 1. Comparisons of Floral Patterns and Parts of the Orchid with the Lily

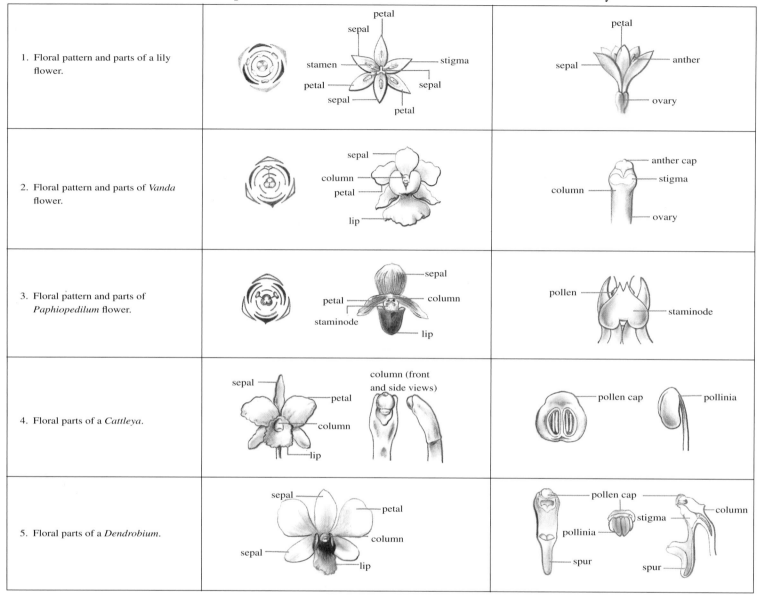

1. Floral pattern and parts of a lily flower.		
2. Floral pattern and parts of *Vanda* flower.		
3. Floral pattern and parts of *Paphiopedilum* flower.		
4. Floral parts of a *Cattleya*.		
5. Floral parts of a *Dendrobium*.		

wavy and decorated with frills at the edge. At the throat there is usually a splotch of yellow. The tip of the column can be seen peeking through the throat, and the sidelobes of the lip have to be removed to display the column. Nevertheless, this inconspicuous structure is the most consistent characteristic of all orchids. The column is slightly curved and expanded towards its tip, bearing on top the pollen grains which have been united into four yellowish masses known as pollinia. The anther has two cells, each containing a pair of pollinia. The pollinia are covered by a loose anther cap. In front, the column tapers to a beak called the rostellum and below this is the hollowed, sticky cavity of the stigma.

The orchid which started the orchid industry in Southeast Asia and later gave rise

Opposite, Clockwise from Top: *V.* Miss Joaquim; *Epidendrum cochleatum* (in this genus, the lip occupies a superior position); *Dendrobium anosmum*

to commercial interest in orchids in Hawaii is the *V.* Miss Joaquim, an extremely free-flowering hybrid which is resplendent in full sun. In the flower of *V.* Miss Joaquim, the petals are twisted around so that the colourful surface that is facing front is actually the back of the petals. The lip has a curious pouch called the spur which points downwards and backwards. If you look at the young buds at the top of the inflorescence, you will notice that the spurs are pointing upwards. Lower down, the larger buds can be seen to have twisted around at the base of the ovary so that in some the spur is pointing sideways and in the oldest buds they are pointing downwards. This semicircular, 180-degree twist occurs in most orchids, but in the *Epidendrum* the cruciate lip remains at the top of the flower. The petals and sepals of the *Epidendrum* are similar and the word 'tepals' has been coined to refer to sepals and petals when they are identical in appearance.

Let us examine the *Dendrobium*. Again, there is a pouch-like structure behind the lip but this is now formed by the fusion of the lateral sepals to a downward extension from the column known as the column foot. The pointed pouch is called the mentum, or chin. The column is very short but has a large stigmatic cavity and on top there are four pollinia hidden beneath the anther cap.

The Slipper Orchids (*Paphiopedilum* and *Cypripedium*) belong to the *Pleonandrae* which have more than one anther. A representative of this group is the *Paphiopedilum barbatum* which still grows in the hills above my hometown of Penang, Malaysia. The plant bears a single flower at the end of the flower stalk and, at first glance, the flower appears very different from the other orchids which we have examined. The dorsal sepal is round and greatly expanded. The lateral sepals are not visible in front, but on turning the flower backwards we can see that these have been

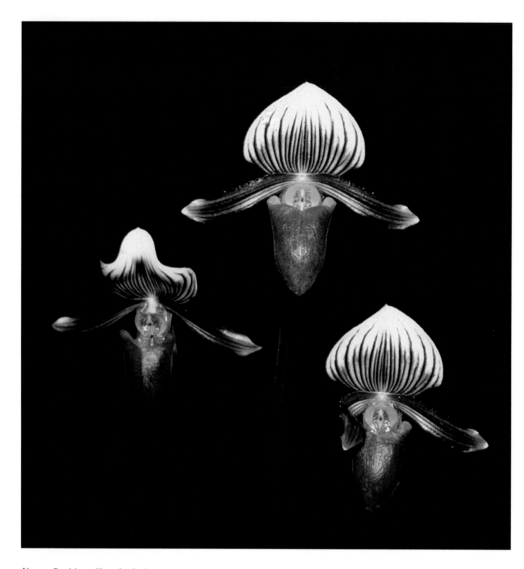

Above: *Paphiopedilum barbatum*

Opposite: *Blc.* Greenwich 'Elmhurst' AM/AOS is a sympodial orchid with pseudobulbs. This plant is grown on an artificial fibreglass rock next to a waterfall.

lip and is occupied by a shield-like structure called the staminode which represents the sterile stamen corresponding to the functional stamen of the *Monandrae*. On either side of the column, behind the staminode, are the two functional stamens, each of which carries two masses of sticky pollinia that can be removed easily. On the underside of the column is a large stigma.

To summarise, orchids may be distinguished from other flowering plants by the following floral characteristics:

1. The sexual organs of the flower (stamens and stigma) are incorporated into a single structure called the column.
2. The anthers are reduced in number to one or two and the pollen are massed into waxy structures known as pollinia.
3. One of the petals has evolved into a highly complex lip.
4. The ovary is inferior and is generally twisted through 180 degrees.

The Orchid Plant

On the basis of their vegetative structure and pattern of growth, orchids are classified as monopodial or sympodial. The monopodial orchid plant has a single stem which produces new leaves at its growing tip and roots lower down. The sympodial ('feet together') orchids grow as clumps of plants whose shoots ('feet') are linked together at their base. The *Cattleya* reminds us of the ginger plant which is a rhizome, that is, it has a creeping stem which gives off vertical shoots bearing leaves and flowers. The *Cattleya* is a sympodial orchid. The basic structural unit of the *Cattleya* is the pseudobulb, a tough, swollen stem whose main function is to store food and water for the plant. It is different from the true bulbs (lily or onion bulbs) which consist of fleshy, scale-like leaves. The larger, unifoliate *Cattleya*

fused together in the midline to form the synsepalum (*syn* in Greek meaning 'together') which is hidden behind the massive lip. The petals spread out on either side of the dorsal sepal and they are covered with hair at the edges and dotted with a few black warts. The lip is in the form of a slipper, earning the orchid its generic name, *Paphiopedilum*.

(*pedilon* is Greek for 'slipper', and Paphos is a town in Cyprus, the favourite island of Aphrodite, celebrated for its temple devoted to the goddess of beauty.)

The column of *Paphiopedilum barbatum* is very different from the column of orchids belonging to the *Monandrae*. The end of the column points forwards at the base of the

has a single leaf to each pseudobulb and this is waxy and thickened to conserve water for the plant, enabling it to thrive as an epiphyte. *Cattleya* plants which have smaller flowers usually have two leaves to each pseudobulb and they are called bifoliate *Cattleya*. In a strong, young pseudobulb, you can see the floral sheath pushing upwards through the lip of the pseudobulb at the axil of the leaf. The floral sheath is a thin, pale green envelope which covers and protects the developing buds. When the plant is ready to flower, the developing buds can be felt within the sheath and the flowers open over the next 21 days. In old pseudobulbs, the sheaths become thin and papery.

When the pseudobulb finishes flowering, a vegetative bud at the base of the pseudobulb starts to swell and breaks through its thin covering scale. It grows horizontally for 2–3 cm, before turning upwards to expand into a new pseudobulb. At its feet, it produces tough new roots which anchor the plant to its support. Old bulbs almost never produce new roots. As its base, the pseudobulb also carries new buds which remain dormant for some time. There are two dormant buds to each pseudobulb and usually one will continue the growth of the rhizome after flowering. However, if the plant is strong, both buds may develop at the same time. The dormant buds of old pseudobulbs can be stimulated to develop when there is a break in the continuity of the rhizome. If each pseudobulb is separated from the others all along the rhizome, each will produce new leads, but the new pseudobulb will be weak and growers usually cut the rhizome at every third pseudobulb when they wish to propagate the *Cattleya*. Old pseudobulbs remain alive for many years and the 'eyes' from back bulbs can be stimulated to develop new plants later on.

Pseudobulbs come in all sorts of shapes. They are long vertical canes up to 1.5 m in

length in the horn *Dendrobium*, dangling pendulous rods in the nobile *Dendrobium*, round and fat in terrestrial orchids such as the *Epidendrum* and *Spathoglottis*, flat and wrinkled in the mature bulbs of *Oncidium*. The presence of storage bulbs enables the plant to withstand periods of drought. In the laboratory, bulbous orchids have remained alive without watering for up to eight years. This feature enabled the epiphytic orchid to survive growing suspended on the tall canopy of the jungle, and terrestrial orchids to establish on scrubland as early pioneers. In some orchids, the pseudobulb burrows deep into the ground. Such plants are capable of surviving bush fires, flowering with renewed vigour after such events.

The Venus Slipper Orchids (*Paphiopedilum*) grow on leaf mould in the deep shade of the forest or in the bog (*Cypripedium*) and, being kept constantly moist, they have not developed pseudobulbs. They are sympodial in their growth pattern and each new shoot arises from a bud at the base of a mature plant. The stem is short and bears two rows of leaves which are close together. At its base, it sends out thick roots which spread horizontally.

Monopodial orchids do not have pseudobulbs and exist as single-stemmed individual plants. They are found in the tropical and subtropical regions of Asia, Africa and Australasia where they grow as epiphytes. The finest examples of monopodial orchids are the majestic strap-leaf *Vanda*. The single stem is stout and bears two rows of alternating, thick, leathery, horizontal leaves. Roots arise perpendicular to the leaves. They grow horizontally at first; then, because of their weight, they bend downwards. Whenever they come into contact with a solid support, they will grow along its surface and become attached by fine root hairs. Water tropism appears to exert a dominant influence in the direction of root growth.

When the plant grows tall and lanky, the stem becomes top-heavy and is easily bent or broken by strong wind. Below each point of rupture in the conducting system of the stem, new shoots appear at the leaf axils, growing into young offshoots possessing leaves and roots. These plantlets are known as 'children', *keikis* to the Hawaiians and the rest of the United States, and *anaks* in the Malaysian and Indonesian regions. *Anaks* are commonly produced in clumps at the base of old plants, particularly those which have lost their lower leaves.

The strap-leaf *Vanda* is a slow-growing plant but many monopodial orchids such as the terete-leaf *Vanda* (for example, the *V.* Miss Joaquim), the curious Scorpion Orchids and the spectacular red *Renanthera* grow very rapidly. They need to raise their heads to the sun to flower and will scramble up trees to which they are attached to get into open sunlight.

At the other extreme, we have the

Phalaenopsis, whose short stem is completely hidden by the bases of its large leaves. Long or short, the stems of monopodial orchids store up sufficient nourishment to enable old, leafless and even rootless stems to put out new offshoots, provided they have not been destroyed by rot.

Water Conservation

Water conservation is a priority for all epiphytes, and a unique characteristic of epiphytic orchids is the root. It is usually fleshy and cylindrical, or flattish as in some *Phalaenopsis*. When the roots are dry, they appear silvery white on the outside and only the growing tip is green or purple. When wet, the entire root turns green. The silvery appearance is due to a spongy coating of dead cells called the velamen, which constitutes an impregnable layer protecting the conducting systems of the root from water loss through transpiration and evaporation.

These pictures illustrate the leaf types of monopodial orchids. The terete leaf is well adapted to withstand dehydration and all terete to quarter-terete *Vanda* thrive in full sun.

Opposite, Left to Right: *Ren.* Tom Thumb x *storiei*; *Aerides* Amy Ede, with strap leaves; *Luisa* hybrid, with terete leaves

Clockwise from Top Left: *V.* Lily Wong, a semi-terete *Vanda*; *V.* Nonito Dolera, a quarter-terete *Vanda*; *V.* Ruby Prince, a three-quarter terete *Vanda*

Water falling on the root is absorbed by both the velamen and the root tip, but only the latter conducts the water into the plant. In its natural epiphytic state, the velamen traps the first run-off from the bark which contains the highest concentration of dissolved minerals, and by subsequent leaching it continues to supply nutrients to the root tips when the run-off water is deficient in minerals. It has been found that there is a high correlation between the thickness of the velamen and the severity of the environment. Whereas moist-growing species have only a single cell layer of velamen, species growing during severe dry seasons or in exposed areas have up to seven layers of cells in the velamen.

When a root first appears on the stem, it is green, and as it grows longer the velamen develops behind the green tip. The roots are capable of varied responses. As soon as they touch a firm support, they grow around it, producing root hairs that cling to the surface. At the point of contact with the support, the velamen is thin or absent and absorption of liquids occurs freely. In climbing monopodial orchids, the lower roots sink into the ground and produce an extensive branching network of cream-white, 'naked' feeding roots, devoid of velamen. These feeding roots are always associated with a fungus, the mycorrhiza, which assists in the breakdown of organic matter in leaf mould and ground debris. Any root can become a feeding root if it finds a suitable substrate. If a ground cover of leaf mould or wood shavings is placed at the base of *Vanda* and Scorpion Orchids tied to wooden stakes, feeding roots will sprout at the base of the stems. In potted plants, the feeding roots develop within the pot under the surface of the potting medium. When I tried growing *Phalaenopsis* in a hydroponic (Luwasa) system, I discovered that enormous plants were being supported by only a single feeding root.

In a few Malaysian orchid species,

Taeniophyllum and *Microcoelia*, the roots have taken over the function of photosynthesis and the short stem is devoid of leaves. Australia, Malaysia and China have several genera of saprophytic orchids which grow in the ground in association with fungi on which they are completely dependent for their food supply. These orchids have no green leaves and do not possess any chloroplast at all. The unusual Australian orchid, *Rhizenthella gardeneii*, even produces its flowers and seeds completely underground.

Two species of moth orchid, *Phalaenopsis stuartiana* and *Phalaenopsis schilleriana*, produce plantlets at their root tips. In the laboratory, orchids have also been cloned from root tips.

Opposite: *Den. cretaceum*

Above, Left: *Den. uniflorum*

Above, Right: Yamamoto *Dendrobium*

The two evergreen *Dendrobium* (*Den. cretaceum* and *Den. uniflorum*) are native to the equatorial monsoon belt which experiences rainfall throughout the year. The plants retain their leaves throughout the year. The Yamamoto *Dendrobium* were built on species which are native to northern Myanmar and Thailand which experience a distinct dry season in winter. These plants must go through a dry spell and totally shed their leaves before they will flower.

Chapter 3
Life History of the Orchid

New conservation laws that are now enforced by many governments prohibit the wide-scale collection of orchid species from their native habitat, their export from their country of origin and their import into the affluent orchid-growing areas. This is in marked contrast to the situation prevailing in the last century, when orchid hunters roamed the tropical regions of the world in search of new species; they collected plants by the thousands and invariably stripped jungles of their orchid treasures to meet the demands of the incredibly insatiable importers in Europe. There were two reasons for collecting the species in such large numbers: after a long ocean voyage, few of the plants reached Europe in a healthy state, and nobody at that time had fathomed the secrets of orchid seed germination. In nature, the orchids grew, miraculously, by the thousands on tree branches; on the ground, they even colonised newly cleared land. Yet all attempts to germinate their seeds in greenhouses met with failure.

In 1899, a French botanist, Noel Barnard, made a monumental discovery. He discovered that the seeds of a terrestrial orchid, *Neottia*, had germinated in some fruits which had lodged under fallen leaves and which were infected by mould. He was quick to realise

Opposite: *Bulbophyllum echinolabium* has the largest flower in the entire genus.

Right: *Den.* (Elizabeth x Tuan Pink). A good white *Dendrobium* with broad, twisted petal would be very desirable. This Singapore Botanic Gardens hybrid is an example of the breeding towards that goal.

the significance of the mould and managed in 1903 to germinate the seeds of a hybrid *Laeliocattleya* in a pure culture of fungus collected from the roots of the pod parent. In a concoction prepared from the ground tubers of the orchid (*salep*), Barnard managed to germinate the seeds of *Bletilla hyacinthina*, a terrestrial orchid, and then in a mixture of *salep* and sucrose he successfully germinated the seeds of *Laelia*, an epiphytic orchid. His work laid the foundation for the next important breakthrough by the American scientist Dr. Lewis Knudson, who concluded that the only function of the fungus was to provide simple sugars for the germinating orchid seeds. Knudson prepared a mixture of inorganic nutrients (fertiliser), added cane sugar and solidified it with agar to produce a growing substrate similar to that employed by medical microbiologists for growing bacteria and other micro-organisms. When he sowed the orchid seeds under aseptic conditions on this agar, they germinated and grew into sturdy plants. Today, millions of seedlings of an orchid species can be produced from a single seed pod in Knudson's C medium.

Orchid seeds are mostly minute structures produced by the thousands or even millions per seed pod, depending on the genus. Each seed weighs between 0.3 and 0.6 microgrammes and all are totally devoid of storage or nutrient tissue. For the orchid seed in nature, finding a fertile substrate with the appropriate fungus is purely fortuitous. With these one-in-a-thousand odds, the otherwise seemingly wasteful production of millions of seeds is in fact essential to the survival of the orchid.

Charles Darwin noted that a single plant of the European Bee Orchid, *Ophrys maculata*, produced enough seed in one season to cover an acre. An acre of orchid would produce 30,317,760,000 seeds the next season, assuming that all the seeds germinated and all the seedlings reached maturity. If this continued unabated, the great grandchildren of a single plant would cover nearly the entire globe.

To gear itself for the production of millions of seeds in a single pod, the orchid flower has concentrated its pollen into two, four or eight pollen masses or pollinia so that fertilisation is a hit-or-miss affair. In many instances, elaborate contrivances are made to attract, ensnare and deposit pollinia on insect pollinators. These include the use of bold colour, including dazzling white; patterns on the lip of the flower; grooves to lead the insects towards the reproductive structures; pouches, springs, projectile mechanisms; insect mimicry; scent

On these two pages, from Left to Right: *Catasetum pileatum*, male flower; *Cycnoches ventricosum*, male flower; *Cycnoches ergotonianum* var. *aureum*. The Central American *Catasetum* and *Cycnoches* have developed very elaborate mechanisms for pollination by bees. Male and female flowers of many species of *Catasetum* and *Cycnoches* are even dissimilar; *Coelogyne rochussenii*. Gregarious flowering is the route to cross-pollination in many Malaysian species.

(Photos on these two pages by courtesy of John Teoh)

and nectar. In the mid-19th century, there was great interest in the mechanisms of orchid pollination as a result of the publication of Darwin's *The Various Contrivances by which Orchids are Fertilised by Insects*. In this book, he postulated that the white 'Star of Bethlehem' orchid, *Angraecum sesquipedale*, which had a spur about a foot (30 cm) long with the nectar filling only an inch and a half (4 cm) of the long nectary, must be pollinated by "moths (in Madagascar) with probicides capable of extension to a length of between 10–11 inches (25.5–28 cm). This belief of mine has been ridiculed by some entomologists…" *Xanthopan morgani praedicta* is the name given to the moth predicted by Charles Darwin 40 years before it was actually discovered on the island of Madagascar.

Most white orchids are pollinated by night-flying moths, colourful orchids usually by bees and smaller orchids by mosquitoes. The famous Hawaiian hybridiser Goodale Moir, creator of some of the most beautiful and difficult hybrids of *Oncidinae*, observed that difficult intergeneric crosses seem to be made in greater numbers on the six days preceding the new moon or the full moon. Perhaps the orchid flowers have adapted to the biological rhythms

of the night-flying insects. The terete *Vanda* of Malaysia and Singapore are pollinated by large carpenter bees (*Xylocapa*). The insects are not particular about which flowers they visit, and in communities where closely related species grow together, cross-fertilisation sometimes occurs, resulting in the appearance of natural hybrids. Dr. Caloway Dodson of the University of Miami, a leading authority on the subject of orchid pollination, calculated that about 2,000 species of orchids are pollinated by the *Euglossine* bees. He made the fascinating discovery, during his honeymoon, that the bees which he had been studying were as attracted to his wife's perfume, Diorissimo, as to the flowers of *Stanhopea tricornis* which had an identical scent from benzyl acetate.

Individual orchid scents are due to any one of 50 oily compounds which are produced by highly specialised scent glands located on the lip. It may be cineole, the scented oil in Vicks VapoRub. The commercial vanilla essence, vanillin, is obtained from the seed pods of the orchid *Vanilla*; it is also produced by the flowers of *Dendrobium leonis*.

Not all orchid scents are pleasing. *Arachnis flos-aeris insignis* (syn. *Arachnis moschifera*)

gives off a strong musk (*moschifera* means 'musk-scented') and some *Bulbophyllum* have the most unpleasant odours. When *Bulbophyllum beccarii* first flowered at Kew, it produced such a strong stench that it rapidly emptied the greenhouses of both botanists and visitors.

Once the insect is lured to the flower, colourful markings and ridges on the lip serve as guides to lead the insect towards the nectary, causing the insect's head to come into contact with the stigma. In many orchids, the spring effect of the hinged lip causes the insect to fall into the spur. As it struggles to come out, its head brushes against the rostellum and the pollinia are deposited on it. In the *V.* Miss Joaquim, the two pollinia are attached by a flat strip of tissue called the stipe to a sticky disc, and when the whole structure is dislodged from the flower, the stipe curls backwards near its tip, raising the pollinia which are now held upwards, thus enabling them to meet the stigma of the next flower visited by the bee. In the genus *Catasetum*, the pollinia are flung onto the insect when it lands on the depressed lip. In the case of the Central American *Cycnoches* (Swan Orchid), the bee lands in an inverted position on the lip of the male flower, but as

Above: *Robequetia cerina.* The inflorescence resembles a bunch of grapes. The rounded flowers barely open in this very successful Philippine species.

Opposite: *Dendrobium palpebrae* exudes the scent of apple.

it attempts to reach the nectary, the posterior legs come off, the abdomen swings downwards and touches the column which whips round and strikes the bee, an action which causes the viscidium carrying the pollinia to be pasted on the tip of the bee's abdomen. Eventually the anther cap drops off and the stipe straightens so that the pollinia are better presented. When the bee flies off after feeding from the nectary of a female flower, the pollinia are entangled in the processes extending from the column.

In the Bucket Orchid, *Coryanthes trifoliata*, the clumsy insect falls into a pool of liquid secreted by the glands. The only way for it to escape drowning is to crawl out between the column and the lip, inevitably picking up the pollinia on the way.

Perhaps the most novel method is the one used by the terrestrial Mediterranean orchid, *Ophrys cretica* — pollination through pseudo-copulation. This particular deception is complex, involving smell, sight and touch. The flowers produce a scent resembling the sexual odour (pheromone) of the bee. When the insect approaches the flower, the latter's resemblance to the female stimulates the insect to attempt copulation. Finally, as it alights on the flower, contact with the hairs on the lip triggers off strong copulatory movements in the insect and pollen transfer takes place.

Flowers in general usually lure insects with the promise of food. Some nectarless orchids, like the equitant *Oncidium*, *Epidendrum radicans* and *Orchis israelitica*, mimic flowers which are rich in oils or nectar that are preferred by bees. The Malaysian *Bulbophyllums*, which are pollinated by bluebottle flies, emit the smell of rotting carrion throughout the day. To protect their pollen, some orchids offer pollen-like markings on their lip for the insects to chew on. Another Mediterranean orchid, *Epipactis consimilis*, resembles aphids in shape and colour and it is attacked by hoverflies which feed on aphids. Some *Oncidium* species are

attacked by territory-defending *Centris* bees.

In spite of all these contrivances, few orchid flowers ever get pollinated (perhaps one in a thousand blooms). Some orchids do away with the need for pollinators, and if they are not cross-pollinated, they will self-pollinate (as seen in *Dendrobium mirbelianum*).

Following pollination, the petals and sepals fade, the stigma closes up and the ovary (seed pod) starts to swell. The development of the fruit capsule is a long drawn-out process, taking 4–5 months in *Dendrobium*, nine months in *Cymbidium* and over a year in some *Vanda* because the growth of the pollen tubes is a slow process with a greater lapse of time between pollination and fertilisation in the more advanced species. In *Habenaria*, the pollination-fertilisation interval is only 8–10 days; in *Spathoglottis plicata*, 15 days; in *Dendrobium*, 8–10 weeks; in *Paphiopedilum*, 3–5 months; in *Phalaenopsis*, three months; in *Vanda tricolor*, 5–6 months; and in *Vanda sauvis*, 6–10 months. Despite the apparently slow speed of seed pod development, it is an energetic metabolic process requiring stimulation by auxins, mainly NAA (naphthalene acetic acid) from viable pollen. In some *Phalaenopsis* flowers, such as *Phal. violacea*, *Phal. lueddemanniana*, *Phal. amboinensis* and *Phal. mariae*, instead of fading, the petals and sepals become green and fleshy, providing additional photosynthetic tissue to support pod growth. Photosynthetic cells on the covering of the seed pod and along the inflorescence also play a role. I once pollinated a plant, *Phalaenopsis* Jimmy Hall, only to have the entire plant killed by bacterial rot four weeks after the pollination, leaving just the green flower spike which carried the developing green pod. Although the inflorescence began to dry up from below, the pod held, and two months later, I was able to obtain viable seed from the crossing.

In the laboratory, it has been possible to culture the ovaries from the cut flowers of *Dendrobium nobile* in nutrient solutions fortified by 6 percent maltose, peptone and coconut water and later to get viable seeds from such ovaries. It worked equally well with *Dendrobium* Jaquelyn Thomas when the ovaries were planted upside down with the stigmatic end inserted into agar medium that contained NAA.

It is interesting to note that although orchid species come in diverse forms and coloration, their fruit capsules are remarkably similar. Initially, they are green but as the pods mature they gradually turn yellow. Hybridisers in the tropics have to spot this change and harvest quickly for, in hot weather, the pods will burst open a few days after turning yellow, and the seeds will be lost or become contaminated with fungal spores.

The orchid pod breaks in a single longitudinal split at first, but as it dries and contracts, either two or five additional splits develop. Seed counts have shown that *Cattleya aurantiaca* produce 256,000 seeds in a single pod; *Cattleya labiata*, 2–3 million with 929,000 viable seeds; and *Cynoches chlorochilon*, 3,770,000 seeds. Some species, such as *Orchis maculata*, produce only 6,200 seeds.

The light, powdery, near-microscopic seeds become wind-borne and may be carried by air currents for hundreds of kilometres. Because desiccated orchid seeds remain viable for over a year (and when maintained at cool temperatures, they may even germinate after keeping for 10 years), it has been postulated that if the seeds were caught by major air currents and lifted up a few thousand metres, they could drift for years and remain viable in the cool, dry air, germinating when dropped into a suitable environment. Nevertheless, orchid species have rather restricted distributions, and most genera are restricted to a single continental land mass.

Seeds of terrestrial orchids have great buoyancy. They germinate more readily

after pre-soaking in water for 15–45 days, suggesting that, for medium range distribution of terrestrial orchids, water transport may play an important role.

Orchid seeds are ellipsoidal and consist of densely protoplasmic cells containing multiple fat droplets, their total food store since they are devoid of endosperm. The distal end contains smaller cells than the basal end and, in some species, the base ends up in a pointed structure called the suspensor. The entire seed is covered by a seed coat.

The first germination processes are induced by the presence of moisture and are independent of infection by mycorrhiza. In the ellipsoidal embryo, the cells at the bottom end lose their fat vacuoles and enlarge. The embryo becomes pear-shaped or globular and stretches the seed coat. Infection by mycorrhiza, usually a species of *Rhizoctonia*, occurs through the dead suspensor cells, and the mycelia penetrate the dead basal cells of the embryo. The fungus produces an enzyme, amylase, which breaks down starch in the substrate, converting it into sucrose which can be utilised by the orchid embryo. Deeper inside the embryo, the fungi begin storage; and further inside, the storage fungi undergo digestion (phagocytosis) within the orchid cells.

As the seedling enlarges, the seed coat splits lengthwise and the embryo turns green through the acquisition of chlorophyll. Groups of papillae develop on the bottom half of the embryo. They are fine white hair quite visible to the naked eye and gradually reached by the fungal mycelia inside the embryo. The embryo is now 2–5 mm in diameter and is called a protocorm.

Protocorms may be smooth or irregular and some may produce multiple buds on their surface. This is commonly seen in flask culture if the concentration of growth stimulants, such as coconut water, is excessive. The development of several secondary protocorms from the epidermal cells of the initial protocorm, however, does not only occur in asymbiotic culture, it has also been observed by Barnard (1909) in *Vanda tricolor* and other hybrids through the action of *Rhizoctonia*. In the normal course of events, a ridge develops on the apex of the protocorm, and from this emerges the first leaf. After several leaves have appeared, roots develop from the bottom or the side of the protocorm.

The fungal host relationship which is laid down in the young seedlings persists throughout the life of the orchid. In the outermost layer, fungal mycelia continually infest new cells. Further inside, surrounding the vascular layers, phagocytosis occurs. In roots, fungus is present only on the basal surface, where the roots are in contact with the substrate.

Once green leaves are produced and strong roots have safely anchored the orchid on its perch, further development proceeds in the normal way following the patterns of monopodial and sympodial growth described in Chapter 2. Whether the seedling is successful now depends on its adaptability to its environment. Epiphytic orchids are early settlers following upon algae, lichens, liver-worts and mosses, and apart from ferns (and in Central America, bromeliads), they do not encounter much competition. In older forests, the orchids select the higher reaches of the forest canopy, where there is too much light and too little moisture for the ferns and aeroids. Trees growing along river banks are also exposed to more sunlight and are favourite hosts for many orchids. Many terrestrial orchids are adapted to disturbed areas, and some grow on cliff surfaces or in rock crevices where few other plants would thrive.

Right: The natural high-altitude, tropical forest habitat has been cleverly simulated in the Cool House of the Singapore Botanic Gardens.

Chapter 4

Sympodial Orchids

The process of description and classification (taxonomy) of plants was pioneered by Carl Linnaeus of Uppsala. He grouped or separated the plants on the basis of their sexual characteristics. Linnaeus employed a binary system for describing plants, each name consisting of:

1. The name of the genus, and
2. The name of the species; as it were, surname first, personal name second. For example, *Vanda* (genus) *hookeriana* (species).

In his *Species Plantarum*, published in 1743, Linnaeus described 60 species of orchids grouped into eight genera. Today, there are over 17,000 species and some 750–800 genera recorded in the herbarium folders at Kew Gardens, England.

E. Pfitzer next divided the orchid family into two subfamilies by counting the anthers: in the *Monandrae*, the flowers have one anther, while in the *Pleonandrae*, there are two or three anthers. The subfamilies are then further separated into tribes, genera, sections, species, varieties and clones. Over the years, various taxonomists have proposed more subfamilies which were created by grouping together those tribes which may have common ancestors. From time to time, taxonomists also resurrect old names that have priority, and create new ones when they think there are sufficient differences to constitute a distinct genus. For example, in the genus *Vanda*, the terete *Vanda* are given the genus name *Papilionanthe*; and Holttum had, long ago, isolated *Vanda sanderiana* as a separate distinct genus called *Vandanthe*. Similarly, a section of *Bulbophyllum*

could be distinguished and given the genus name *Cirrhopetalum*. However, within the academic botanic community there is no common agreement on the entirety of any of these proposals. All these create considerable confusion for the layman.

I have chosen to stick to my earlier decision to adopt the proposal of Frederick Gustav Brieger at the 8th World Orchid Conference in Frankfurt (1975) to maintain a consistency with the earlier editions. The five families are the *Cypripediodeae*, the *Neottioideae*, the *Orchidioideae*, the *Epidendroideae* and the *Vandoideae*, the first three showing primitive forms and the last two being highly evolved forms.

The subfamily *Cypripedioideae*, sometimes referred to as *Pleonandrae*, consists of the *Cypripedium* tribe, whose members have a pouch-shaped lip and two anthers (hence their popular names Slipper Orchid or Moccasin Orchid), and the *Apostasia* tribe (possessing two anthers) and the *Neuwiedia* tribe (with three anthers), both of which do not have the pouch-shaped lip. The *Apostasia* tribe is native to the Indo-Malaysian region but it is uncommon and devoid of horticultural interest so it will not be discussed.

Among the monandrous orchids (with one anther), *Orchidioideae* is a large subfamily to which the majority of European and African terrestrial orchids belong. The anther does not fall off after the removal of the pollinia. Many members have subterranean tubers which allow them to survive the winter frost. The *Orchidioideae* is represented in Southeast Asia by the single tribe, *Habenaria*, which itself is not commonly cultivated.

Opposite: *Cymbidium wilsonii*

Above: *Dendrobium anosmum* var. *alba*

They are both members of the subfamily *Epidendroideae*.

In the remaining three subfamilies (*Neottioideae*, *Epidendroideae* and *Vandoideae*), the anther falls off when the pollen is removed because it is not physically connected to the rostellum. They belong to a group called the *Acrotonae*, which contains the bulk of the tropical orchids. The *Neottioideae* is a group of primitive, sympodial, terrestrial, herbaceous orchids whose principal characteristic is the presence of granular pollen. They have no tubers. The Jewel Orchids, the saprophytic orchids and the fascinating high-mountain genus *Corybas* belong to this subfamily.

The subfamilies *Epidendroideae* and *Vandoideae* have waxy pollen. They constitute the bulk of the cultivated orchids grown in the tropics. The *Epidendroideae* are sympodial orchids while the *Vandoideae* embrace all the monopodial orchids. Within the *Epidendroideae*, there are the following tribes: *Phaius*, *Nephelaphyllum*, *Eulophia*, *Arundina*, *Bromheadia*, *Liparis*, *Dendrobium*, *Bulbophyllum*, *Coelogyne*, *Cattleya*, *Odontoglossum*, *Acriopsis*, *Cymbidium* and a few others. The *Vandoideae* consists of a single tribe, the *Vanda-Arachnis* tribe (or *Aerides* tribe).

For the sake of simplicity, all the sympodial orchids are placed in one chapter (four subfamilies) and the monopodial orchids in another (one subfamily).

The Naming of Orchids

In keeping with the recommendations of the International Committee on Orchid Nomenclature, natural species are designated by italics (as in *Arachnis hookeriana*) whereas man-made hybrids are designated by plain type (as in *Arachnis* Maggie Oei). Varietal names are in italics if they are naturally occurring clones, such as in *alba* (white), *orchracea* (yellow), *rubra* (red). Personal varieties or awarded clones are designated by plain type, as seen in 'Sagarik strain', 'Merah'. If a specific

plant is known to have received an award, this will be indicated after the varietal name. For example, FCC/RHS (First Class Certificate of the Royal Horticultural Society), AM/AOS (Award of Merit from the American Orchid Society), HCC/OSSEA (Highly Commended Certificate by the Orchid Society of Southeast Asia). In the case of the OSSEA awards, the year of the award is also indicated because standards can be expected to improve over time. Unfortunately, the photographs here only show the orchids as they were seen by the author, on the day he took the photograph; it does not show the condition of the plant or the standard of the flower when it received the award, which should have been better.

The names of all the genera are in italics, regardless of whether they are naturally occurring or man-made.

At one time, the Nomenclature Committee advocated the use of *-ana* to replace *-iana* for the terminal portion of a species name; hence the first edition of this volume used names like *Vanda sanderana* and *Vanda hookerana*. As the Committee has reversed its decision we are back to *Vanda sanderiana*, etc. The separation of *Paraphalaenopsis* from *Phalaenopsis* has created a more difficult problem. Singaporean and Malaysian breeders have done the most work with this genus and there is string of multi-generic names with local flavour (*Trevorara, Sappanara, Laycockara, Starmariaara, Yapaara, Bochoonara, Yeepengara, Waironara, Leeara, Edeara*, etc.) honouring the contribution of the local orchid breeders/growers. Such names are retained in this edition to maintain the proper origin and identity of the hybrids.

These orchids illustrate the sympodial nature of growth and flowering. The inflorescence arises laterally from any node in *Trichostosia* whereas it arises from the apex of the individual plants in the *Paphiopedilum*.

Opposite: *Trichostosia ferox*. The new genus of 50–60 species, which are rarely seen, was split off from *Eria* and is characterised by the short red hair covering the entire plant and flower stems.

Above: *Paph. concolor*

Subfamily: *Cypripedioideae*

The subfamily *Cypripedioideae* comprises four genera: *Paphiopedilum*, *Cypripedium*, *Phragmapedilum* and *Selenepedilum*. All of them are characterised by the pouch-shaped lip and the presence of two anthers. The genus *Paphiopedilum* is the largest, distributed throughout tropical Asia, mostly at high altitudes. It is a paradox that, while the imported *Paphiopedilum* is an important horticultural genus in the temperate region with which thousands of hybrids have been made, it is decidedly uncommon in orchid collections in the Asian tropics, because the cities are at sea level, and the *Paphiopedilum* simply will not flower properly in the heat and do not need the intense light. It is time that growers in Southeast Asia, the natural home to so many *Paphiopedilums*, learn to cultivate this beautiful genus.

Cypripedium are the Lady Slippers of the northern temperate zone. The moccasin flower of Minnesota (*Cypripedium reginae*) also occurs in China. *Phragmapedilum* and *Selenepedilum* are natives of South America, the former genus being easily recognisable by the long trailing petals. These three genera are not commonly cultivated, unlike the 'Paphs'.

Genus: *Paphiopedilum*

Paphiopedilum, 'Paphs' for short, is a genus of terrestrial orchids with some 70 species which are native to South and Southeast Asia, with a distribution that extends from the Himalayas and Myanmar into Indochina and the Asean region up to Papua New Guinea. They were called *Cypripedium* in the past, but botanically *Cypripedium* is a different, although related, genus inhabiting the temperate regions of the northern hemisphere, and present-day horticulturalists appreciate the differences between the two genera. 'Paphs' are popularly known as the Venus Slipper Orchids, Lady Slipper Orchids or simply Slipper Orchids, because of the resemblance of the pouch-shaped lip to a lady's slipper. In its native Malaya, *Paph. barbatum* is known as the *bunga kasut* ('shoe flower') among the people of Malacca.

The flower structure of *Paphiopedilum* is quite different from the other genera that we have described so far. The dorsal sepal is large and erect. The lateral sepals are fused in the midline to form the synsepalum. The two petals commonly point downwards instead of horizontally or upwards, and the lower half of the lip is converted into a unique pouch. The reproductive parts are even more unusual. The column carries a disc-shaped staminode behind which are two pollinia, one on either side, while the stigma is situated in its usual position on

the inferior surface of the column. The ovary has only one chamber, and this distinguishes it from the related genus *Phragmapedilum*, which has three chambers in the ovary and which is native to tropical America.

'Paphs' are sympodial orchids, each shoot having a very short stem with large, thick, perennial leaves closely arranged in the shape of a fan. The leaves are erect in most species, though in a few, they grow horizontally. As a rule, cool-growing 'Paphs' have plain green leaves while warm-growing 'Paphs' have mottled leaves of dark and whitish green. The roots are stout, at least 2 mm in diameter, sparsely branched and spread horizontally up to 60 cm away from the shoot. They are brownish, densely covered with fine hair, rather like tobacco pipe cleaners. The flower

stalk is always terminal, usually carrying a single bloom, though in some species it may have 2–12 flowers. The flowers are thick and waxy, with exotic colour combinations (even for orchids), and sometimes they are further embellished by interesting venation on the pouch, striping on the dorsal sepal, callosities and bristly hair on the petals. They last for months if they are not pollinated by insects or by man. Once a shoot has bloomed, it will never produce another flower spike, but soon after flowering, one or more new shoots arise from the base of the old shoot, developing rapidly, growing straight upwards.

'Paphs' grow on leaf mould and other organic debris that collect on the ground and in rock crevices. Some species, like *Paph. niveum*, grow on limestone formations overhanging the sea. The Malaysian *Paph. lowii* is epiphytic on old trees. Most species are found at high elevations in the tropics where it is always cool and humid. They are associated with mosses and ferns. These montane species adapt well to the temperate regions, but lack vigour when they are cultivated in the tropical lowlands and do not flower well, if at all. The flowers are stunted when such plants bloom in open cultivation in Singapore. A cool house is the solution, but this is expensive to maintain. However, 'Paphs' require very little light, between 1,000 and 1,500 foot-candles, and cool houses can be shaded to reduce the cooling cost. In the former Temperate House at the Singapore Botanic Gardens, the 'Paphs' bloom in abundance throughout the year.

Opposite: *Paph. micranthum*

Left: Back view of a flower of *Paph. concolor*

Note that the lower sepals are fused at their medial border to form a single structure called the synsepalum.

One reason for the popularity of 'Paphs' is the fact that they are compact plants which require little growing space. Experts have their individual recipes for the best growing medium, but this generally consists of modifications of the following simple mixture: one part leaf mould, one part tree fern, one part gravel and one part charcoal or an alternative porous medium. They grow nicely in 10–15 cm plastic pots. Some growers add crushed egg shell or marble chips to increase the calcium content in the medium for those species which are partial to limestone.

The genus *Paphiopedilum* was subdivided by Pfitzer into three groups or subgenera on the basis of the differences in the shape of the lip and petals, as follows:

1. The *Brachypetalum* Group. The typical member is *Paph. niveum*. This group is native to Southeast Asia. The characteristic features of the group are a large pouch with an in-growing rim and an absence of auricles on the sides, large, broad, round petals and mottled leaves that are purplish underneath. Apart from *Paph. bellatulum*, the rest do reasonably well in the lowlands.

2. The *Anotopedilum* Group. Typified by *Paph. philippinense*, this group is easily distinguished by long, narrow, twisted petals, a pouch that occupies only the distal half of the lip and edges that are not incurved. There are no auricles. The leaves are evenly green, long and strapped.

3. The *Otopedilum* Group. This is the largest subgenus and may be represented by *Paph. barbatum*. As its name implies, there are two prominent auricles, or 'ears', at the base of the lip. The pouch occupies half the length of the lip. The petals are a little longer than the sepals but they are not twisted. The leaves are an even green in some species, tessellated in others.

In his revision of the genus, Phillip Cribb adopted J.T. Atwood's 1984 proposal to divide the genus into two subgenera, based on the evidence of their evolutionary development (see Cribb's *The Genus Paphiopedilum* published by The Royal Botanic Gardens in 1987). The two subgenera are *Brachypetalum* and *Paphiopedilum*. The subgenus *Brachypetalum* consists of the members of the old section *Brachypetalum* and the five newly discovered species of the section *Parvisepalum* from China and Vietnam.

The species of all the other groups fall within the subgenus *Paphiopedilum,* which is subdivided into five sections:

1. *Coryopedilum* has nine species: five in Borneo, three in the Philippines and one in Papua New Guinea. Their chromosome number is 26 and their inflorescence bears multiple flowers which are characterised by long, often twisted, petals. *Paph. philippinense* may be regarded as

a typical member. The most admired member is *Paph. rothschildianum*.

2. *Pardalopetalum* is made up of three species with a wide distribution. Its chromosome number is 26 and its inflorescence also bears multiple flowers with long twisted petals. It is separated from *Coryopedilum* by the obcordate (heart-shaped) staminode. *Paph. lowii* is the member most suited for the lowlands, but *Paph. parishii* is more commonly seen at shows.

3. *Cochlopetalum* bears single flowers; or if there are several flowers on an inflorescence, they open successively, not together. Their chromosome number ranges from 30–37. They are represented by *Paph. bullenianum* and *Paph. hookerae*.

4. *Paphiopedilum* bears single flowers and has green, non-tessellated leaves. Their chromosome number is 26. There are seven species of which *Paph. hirsutissimum* is the most typical. There are many stunning *Paphiopedilums* in this section but unfortunately they all require a cool climate.

5. *Barbatum* has 26 member species, with chromosome number, ranging from 28–42. Their leaves are tessellated. The inflorescence is single-flowered and the lip has distinct sidelobes. *Paph. barbatum* is a typical member.

Extensive and systematic hybridisation has produced a wealth of magnificent hybrids in *Paphiopedilum* which possess all the qualities considered desirable for the genus — full, round form and horizontal axis of the petals, good size, clear, contrasting colours or a clean monochrome. Repetition of perfect forms in advanced hybrids has caused modern hybridisers to shift their attention to the old, primary hybrids which were first done a century ago. By repeating the same crosses with better parents, beautiful hybrids have again been produced, their special charm being their unconventional appearance.

Various species of *Paphiopedilum* bloom at different times of the year and several species remain in bloom for many months. Hybrids may have a blooming season which spans the flowering period of their parents. By selecting a good range of species and hybrids, one can have flowers in bloom throughout the year.

Section: *Brachypetalum*

Members of this section are distributed through Myanmar, Thailand, Indochina and the northern part of Peninsular Malaysia, and all grow on or near limestone.

Paphiopedilum bellatulum

This attractive, well-shaped '*Paph*' is native to Myanmar and northern Thailand where it is found growing in the open or in the shade of limestone hills at 1,000–1,800 m. The leaves are a deep green overlaid with blotches of light green; but the undersurface of the leaves are a deep purple. The plant grows close to the ground as a tuft of leaves which vary in size from 18 by 7.5 cm to 25 by 9 cm.

The hairy purple peduncle is so short that the flower almost rests on the ground (or pot). The flower is round, 6–8 cm in diameter, white or cream in colour and heavily spotted with large blotches of maroon. The sepals are almost circular and lend a contrasting background to the lightly speckled, egg-shaped lip.

The species has been cultivated in Singapore and Malaysia for some time, but it is not free-flowering, unless it is grown in a cool house.

Paphiopedilum concolor

Paph. concolor is a member of the *Brachypetalum* group, thus it has mottled leaves and a rather short inflorescence. It is found in Indochina, Myanmar and the southern half of Thailand, and is easy to grow and flower in the lowlands. It grows in rock crevices in the coastal lowlands, below 300 m, but has been found at 1,000 m or higher. Peak flowering is in May.

Left: *Paph. concolor*

Opposite, Top: *Paph. niveum* var. Ang Thong

Opposite, Bottom: *Paph. niveum*

The flowers are cream to peach yellow, dotted with crimson. The petals and the dorsal sepal are round, the flower measuring 6 cm across. Usually one flower and occasionally up to three flowers are borne on the scape. Several forms have been described and named.

Paphiopedilum godefroyae

Distribution of this mottled-leaf species is similar to *Paph. niveum* and *Paph. concolor*, barely above sea level on limestone cliffs. On the birds' nest islands of Thailand, it is exposed to the afternoon sun. The flowers are large and round, 9 cm across, creamy white to light green and spotted with brownish purple. The flower scape is short, bearing one or two flowers from December to July. Holttum thinks that it may be a natural hybrid, perhaps between *Paph. bellatulum* and *Paph. concolor*.

Paphiopedilum niveum

This is the famous orchid of Pulau Langkawi, an island group off the northwestern coast of Peninsular Malaysia. It has also found in Penang and in the Tambilan islands which are west of Borneo. A variety in Thailand, known as *Paph. niveum* var. Ang Thong, comes from the aforementioned birds' nest islands in the Gulf of Thailand. It grows in rock crevices and ledges on limestone rock, almost down to sea level. Over-collecting from Pulau Langkawi has made the species quite scarce.

This plant is small, with mottled leaves only 10 cm long which extend horizontally from the shoot. It produces one or two small, round, white flowers at the end of an erect, hairy, 15 cm scape. The flowers are 5–6 cm across and, in the common variety, the petals and the dorsal sepal are finely spotted towards the base. The pure white form is exquisite. It flowers from December to August, with a peak in April and May. Kamemoto and Sagarik

believe that variety Ang Thong may be a hybrid, perhaps with *Paph. godefroyae*.

Paph. niveum, while not robust, is easy to grow and it flowers in the tropical lowlands. Many hybrids have been made with *Paph. niveum,* particularly with the *alba* strain, to produce delicate, round, sparkling white flowers.

Section: *Parvisepalum*

Several dazzling new species of *Paphiopedilums* were collected from Vietnam and the Yunnan Province of China in the 1980s. These new 'Paphs' resemble the *Cypripediums* in the gross appearance of their flowers, particularly the lip, and some taxonomists suggest that they are bridges between the two genera. The new species include *Paph. delenatii* from Vietnam and *Paph. malipoense*, the yellow *Paph. armeniacum* and the pink *Paph. micranthum* from Yunnan. Their flowers are large and striking, and the floral parts are round. Their generally excellent form has earned many high awards, including First Class Certificates from the judges of both the Royal Horticultural Society and the American Orchid Society. They all belong to a new section designated as *Parvisepalum*. The plants grow on karst (coral) formations, apparently in a rather small area.

In the past, orchidists in tropical Asia could only admire the photographs of the *Paphiopedilums* but had scant hope of owning a flowering plant. Today, as a result of cheap, reliable and trouble-free air freight and the establishment of huge nurseries in northern Thailand, Taiwan, Guangxi and Yunnan, *Paphiopedilums* are readily available and relatively inexpensive in Southeast Asia. Even though they would be difficult to maintain and reflower in Singapore, the longevity of their flowers make a blooming 'Paph' a worthwhile purchase. However, the five species in the section are protected and may still only be admired in photographs.

Paphiopedilum micranthum

I have chosen this lovely species to represent the new type of *Paphiopedilum* from Yunnan and Vietnam because it is the commonest. The five members of the section *Parvisepalum* are *Paph. armeniacum* (which is golden yellow),

Paph. delenatii (pink), *Paph. emersonii* (white with pink flush), *Paph. malipoense* (apple green) and *Paph. micranthum* (pink). They are intermediate in appearance between the *Cypripedium* and the *Paphiopedilum*. *Paph. armeniacum*, *Paph. malipoense* and *Paph. micranthum* are found growing on karst at the Yunnan-Vietnamese border, with the stated altitude of one species at 600–800 m. This rocky soil on which they grow is covered with fossilised coral. It was once a sea bed which was pushed up by the northeastern movement of India against the Asian continent; the area is subjected to periodic earthquakes.

Paph. micranthum was discovered in the limestone hills of southeastern Yunnan in 1951 and reached the orchid growing circle about 20 years ago. The fact that over 35,000 plants were exported through Hong Kong to the United States in 1982 attested to widespread stripping of the plants from their native habitat.

The plants are small, almost miniature. Their flat, mottled leaves which extend horizontally measure only 1.5–2 cm in width and 5–15 cm in length. Single, beautiful light pink flowers are borne on erect, hairy inflorescences which are 10–15 cm tall. Some plants have extremely large flowers and are suspected of being polyploid. The flowering period is early spring, around March to early April.

Section: *Coryopedilum*

Paphiopedilum philippinense

This is a variable species widely distributed throughout the Philippines, growing on limestone boulders in the open down to sea level. It flowers from January to April. It was discovered by J.G. Veitch in 1856, growing on the roots of *Vanda batemanii* (*Vandopsis lissochiloides*) on rocks by the seashore in a small

island of the Philippine archipelago. It should be noted here that *Vandopsis lissochiloides* grows in the shade at the edge of the forest (therefore exposed to good light), but not in open sun.

Paph. roebellinii has larger flowers with longer pendent petals than the type, but it is, nevertheless, classified by Cribb as a variety of the species.

The plant is green and robust, with up to nine leaves each 20–50 cm long on a single plant. The tall, erect inflorescence bears 2–4 flowers. The flowers are generally a yellowish green and characterised by the long, twisted, narrow petals that are bordered by warts and hair at the upper and lower margins of the proximal segment. The dorsal sepal is white-striped with purple, and hairy on its dorsal surface; the synsepalum is white, veined in

On these two pages: *Paph. micranthum* offered at a flower market in Kunming, Yunnan, 2004. The opposite page shows the range of floral forms of the two small stall holders in a flower market in Kunming who had fewer than 10 plants between them.

green. The pouch is shaped like a Spanish helmet, yellowish and lightly striped with green. It flowers from summer to fall.

Paphiopedilum supardii

Also known as *Paph. devogelii*, this new Bornean *Paphiopedilum* is allied to *Paph. rothschildianum* but differs from it in having shorter but broader, half-twisted, greenish petals and are spotted with maroon. The dorsal sepal is similarly coloured by the maroon which appears as stripes. The narrow pouch is yellow suffused with a dark brown and overlaid with purple stripes. The erect 40 cm inflorescence bears three or more flowers at a time.

Section: *Pardalopetalum*

The three species in this section share two features in common with those of the section *Coryopedilum*, namely a chromosome number of 26, and multiple flowers on the inflorescence. The three species are found at the centre of the distribution of *Paphiopedilum*, namely from Myanmar down to Peninsular Malaysia eastwards to the Philippines and Sulawesi.

They are named for the leopard spots on the petals. In Greek, *pardalis* means 'leopard'.

Paphiopedilum parishii

This apple-green species is native to the highlands of Myanmar and northern Thailand where it grows on thick moss on boulders or on the branches of a *Terminalia sp.* at 1,200–1,350 m. The tips of the petals are shaped

Left: *Paph. philippinense*

Opposite, Clockwise from Top: *Paph.* Jogjae, a pre-WWII hybrid which does extremely well in Singapore; *Paph. parishii; Paph. supardii*

like a spoon, by which reason a mono species section, *Mystropetalum*, was defined. However, we have followed the classification of Cribb, who places it in *Pardalopetalum* on account of the leopard spots on the petals.

Section: *Cochlopetalum*

The name is derived from the Greek *kochlos*, which translates as spiral shell. The section is named for its coiled petals which are covered with bristling hair. The handful of species in this section is restricted to Sumatra and Java, which were once connected by a land bridge. They produce many flowers on the inflorescence which open in succession, thus enabling a vigorous plant to remain in bloom for months or even years.

Paphiopedilum liemianum

The flower of this species is similar to those of *Paph. glaucophyllum* and *Paph. chamberlianum*, but the colours are more brilliant. The plants were collected from Gunong Sinabung at 600–1,000 m in northern Sumatra by the famous Indonesian orchidist, Liem Khe Wie. They are easily distinguished from other members of the section by the rows of dark purple spots on the undersurface of the leaves, and on the bracts of the inflorescence.

The purplish 15–20 cm long inflorescence is covered with fine hair and bears several flowers, each 8–9 cm across. The dorsal sepal and synsepalum are green, bordered by a pale yellow band. The narrow petals are nearly horizontal, cream yellow with dark purple spots and margins, folded and pubescent at the edges, and with a single twist at the distal end. The lip is helmet-shaped, light pink to purple, with a pale yellow rim. The green staminode is flushed with dark maroon.

The chromosome count is 32, compared with 34 in *Paph. glaucophyllum* and *Paph. chamberlianum*.

Section: *Paphiopedilum*

This is a group of 10 species of green-leafed *Paphiopedilums* native to the mountains of the Asian mainland (in India, Myanmar, Thailand, southern China and Indochina), the single exception being *Paph. exul*, which is found near the sea. This section contains some extremely beautiful species, such as *Paph. fairrieanum*. Theirs is also the shape which has the most influence on *Paphiopedilum* hybrids in general. The flowers are borne singly.

Left: *Paph. liemianum*

Opposite, Top: *Paph. kalopakingii*

Opposite, Bottom: *Paph. tonsum*

Paphiopedilum exul

This species is found growing in humus-filled rock crevices in full sun or in the shade of *Pandanus* in southern Thailand, down to Krabi, and is not a difficult plant to grow or to flower. Sagarik and Kamemoto say that it flowers from February to May. The greenish yellow flowers are 5 cm across and 6 cm tall. The dorsal sepal is erect, white overlaid with an oval, central, yellow-green patch at its lower half and stippled with purple spots over this greenish area. The petals are almost horizontal, 4–5 cm in length and up to 1.7 cm wide, yellow with median purple lines, undulating at the margins. The lip is a bright yellow.

It resembles *Paph. insigne* but the latter species is found at 1,200–1,500 m in northeastern India.

Paphiopedilum kalopakingii

This delightful 'Paph' from central Borneo was first seen in bloom in the nursery of Liem Khe Wie (alias A. Kalopaking). Jim Comber saw a plant with 14 flowers on a single spike! The flowers are 6–10 cm across; sepals white, with red veins; petals white with red veins; lips ochre, veined and pointed at their apex. The plant has green foliage.

Paphiopedilum stonei

Paph. stonei grows on limestone rocks in Sarawak at 300–500 m. It was discovered by Sir Hugh Low in 1860. According to Holttum, earlier records showed that it was not free-flowering in Singapore. However, new shipments grow well and flower freely.

The plant is large with leathery leaves of even green measuring 30 cm in length. The 60 cm tall scape bears 3–5 flowers, 12 cm tall. The dorsal sepal is heart-shaped, creamy white and marked with brownish black longitudinal lines. The extremely narrow petals curve outwards and downwards and are creamy yellow dotted with reddish brown. There is fine hair on both sides of the petals close to its attachment to the column. The pouch is cream coloured with a rosy blush in front, and is decorated by a reticulation of maroon veins. Several flowers are open together, producing a very fine show when the plant is in bloom in midsummer.

Paphiopedilum tonsum

Paph. tonsum (*tonsum* means 'shorn') was named by Reichenbach for its petals, which lack the hairy margin common to the other members of the section *Barbata*. It is yellowish green in colour. The species is common in northern Sumatra, growing in damp humus at 500–1,000 m. The flowers are up to 14 cm across. It flowers in autumn.

Paphiopedilum venustum

This species with dark green leaves, 12.5–25 cm by 2.5–5 cm in width, is native to Nepal and Assam, India. It is instantly recognisable due to the appearance of the curious, orange-yellow pouch which resembles a human brain on account of its green veins. The dorsal sepal is white and veined in dark green. The narrow petals are spathulate, white, veined in green and flushed with pink to brownish pink at the tips. They are edged with fine hair and marked with black warts over the superior margin and axis.

Hybrids of *Paphiopedilum*

Hybridisation in *Paphiopedilum* has reached an extremely advanced stage in Europe and the United States with modern, complex hybrids invariably turning up with perfect form,

colour, size and substance. Polyploidy has played an important role in the advancement. It is interesting to note that while triploids and aneuploids are often ignored in breeding work, the triploid *Paph. insigne* 'Harefield Hall' occurs in the background of many modern tetraploid *Paphiopedilum*. The aneuploid (hypotetraploid) *Paph.* F.C. Puddle FCC/RHS is the parent of many outstanding white hybrids. Aneuploidy is common in hybrid *Paphiopedilum* because, unlike *Vanda*, where the genome number is constant (19), there is a variation in chromosome numbers (26, 28,

32, 34, 36, 38, 42) in the various species of *Paphiopedilum*.

The current trend in *Paphiopedilum* is to repeat the old primary and simple secondary hybrids using superior parents. Their non-conformity lends such hybrids special charm in a community of regimented perfection. *Paph.* Shireen (*Paph. glaucophyllum* x *Paph. philippinense*) and *Paph.* Jogjae (*Paph. glaucophyllum* x *Paph. praestans*) are two primary crosses which will grow and flower well in Singapore.

The following species occur at 1–500 m above sea level, and such species or their hybrids should be considered by hobbyists who are interested in growing *Paphiopedilums* in the tropical lowlands: *Paph. concolor*, *Paph. godefroyae*, *Paph. niveum*, *Paph. philippinense*, *Paph. randsii*, *Paph. sanderianum*, *Paph. kalopakingii*, *Paph. stonei*, *Paph. haynaldianum*, *Paph. glaucophyllum*, *Paph. primulinum*, *Paph. exul*, *Paph. bullenianum*, *Paph. hookerae*, *Paph. barbatum*, *Paph. callosum*,

Above: *Paph. sukhakulii* (common type)

Right: *Paph. sukhakulii* (selected). *Paph. sukhakulii* is not a typical example of the variation that occurs in some species of *Paphiopedilum,* which are sometimes almost sufficient for some taxonomists to ascribe new species. Sometimes, the so-called separate species are just botanic variants; sometimes they are natural hybrids with other species occurring in the same locality. The experts are thus divided into two schools — the 'splitters' and the 'lumpers'. The author has chosen to follow the classification proposed by Phillip Cribb (1987), which most appeals to him. It is based on morphology, cytogenetics and the phylogenetic theory proposed by J.T. Atwood (1984).

Paph. lawrenceanum, Paph. acmodontum and *Paph. urbanianum.* Alternatively, there are many superb, award-quality *Paphiopedilum* hybrids available now as supermarket plants and well worth a try.

Section: *Barbata*

This is a large section with 24 species, all of them possessing tessellated leaves and single-flowered inflorescences. The petals are curved and marked by warts and bristles, especially at the upper border. The lip has prominent, incurved sidelobes. It is also known as the section *Phacopetalum.*

Paphiopedilum barbatum

This is perhaps the commonest *Paphiopedilum* cultivated in the lowlands of Malaysia and Singapore. It used to be abundant on Penang Hill at 700 m, growing in masses on granite rocks in shady locations. On Mount Ophir and Kedah Peak (1,000–1,200 m), it grows in moss by streams in exposed places. Very fine forms have been collected from Kelantan and Trengganu. It also occurs in Thailand.

The plants are dwarfish, with mottled leaves of 10–15 cm length. The flowers are handsome, with a prominent, broad dorsal sepal which is greenish white, striped by longitudinal purple veins that fade off before they reach the margin. The petals are narrow, horizontal but spreading slightly downwards towards the tip. Fine hairy warts dot its upper edge and it is striped by green veins which shade into purple distally. The broad pouch is purple. It resembles *Paph. callosum* but it may be distinguished by the differences in the shape of the dorsal sepal, the shape, markings and position of the petals and the shape of the synsepalum. Cytologically, for *Paph. barbatum,* 2n=38, while *Paph. callosum* has 32 chromosomes. The common chromosome number for *Paphiopedilum* is 26.

Paph. barbatum was discovered by Hugh Cumings in 1838 on Mount Ophir in Peninsular Malaysia, the fourth *Paphiopedilum* to be introduced. It has been extensively cultivated and hybridised in Europe. It grows well in Singapore, in broken brick or charcoal and tree fern, but needs a really cool night temperature to flower well.

Paphiopedilum bullenianum

This species was first discovered in Borneo by Sir Hugh Low, but it also occurs on the slopes of Gunung Panti in Johore and on Pulau Tioman off the coast of Johore in Peninsular Malaysia. It grows in heavy shade on the ground, in a thick layer of porous leaf mould which is continuously damp. It is a dwarf species with mottled leaves 10–15 cm in length.

The 30 cm-tall, erect scape carries one or two flowers, each 7 cm across. The green dorsal sepal is keeled (or rolled inwards) at the distal half and marked with longitudinal veins of darker edge. The pouch is mahogany, fading to white at the toe. It does not flower freely in Singapore, except in a cool house.

Paphiopedilum callosum

This species is very similar to *Paph. barbatum*, but the warts on the petals are longer, nearly 3 mm in diameter, the petals are curved in the form of an 'S' and the synsepalum is acute instead of rounded. The flowers are larger, up to 10 cm across. It is native to Thailand, in the mountainous regions of Chiangmai

and eastern Prachinburi. Strange as it seems, it flowers better in Singapore than *Paph. barbatum*, which occurs at lower elevations.

Paphiopedilum lowii

Unlike its brothers, this species is epiphytic on old trees at 1,000–1,300 m on the Main Range of hills in Peninsular Malaysia. It has also been collected from Borneo and Sumatra. It grows well in the lowlands and will flower in Singapore. The leaves are 30 cm long, slender and uniformly green. The scape carries 2–6 flowers which measure 10–12 cm across. The petals are twice the length of the dorsal sepal, outspread, curving downwards. They are yellow to chartreuse and dotted with brownish

Opposite: *Paph. bullenianum*

Below, Left to Right: *Paph.* Maudiae; *Paph. barbatum*

spots medially, becoming rose to violet distally. The dorsal sepal varies from yellow to green, overlaid with a brown blush and streaked with darker brown proximally. The pouch is brown.

Paphiopedilum sukhakulii

Shaped like a bird with outstretched wings, this striking species was discovered only in 1964 by Prasong Sukhakul in northeastern Thailand. It is found on Mount Phu Luang up to 1,000 m in the Loei Province in loamy soil which is rich in leaf mould, beside shaded streams with low light intensity of 800–900 foot-candles. Its vegetative appearance and distribution are similar to that of *Paph.*

callosum, with which it has been confused when not in flower. The most striking feature of the flower are the broad, horizontal, pointed, green petals which are striped with 6–8 narrow veins of dark green, dotted with brownish red spots and rimmed by fine hair along the edges. The dorsal sepal is erect, round but pointed, on a white base striped by 11–13 greenish veins. The pouch is yellowish green towards the toe shading into coppery brown at the rim. The flower is 10 cm across and borne singly on erect, hairy scapes 10 cm tall.

One of the most beautiful species, *Paph. sukhakulii*, flowers consistently in September and October, and sometimes also in spring.

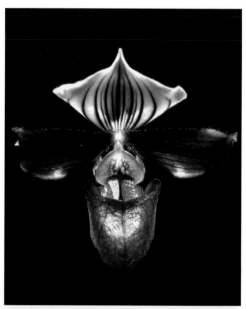

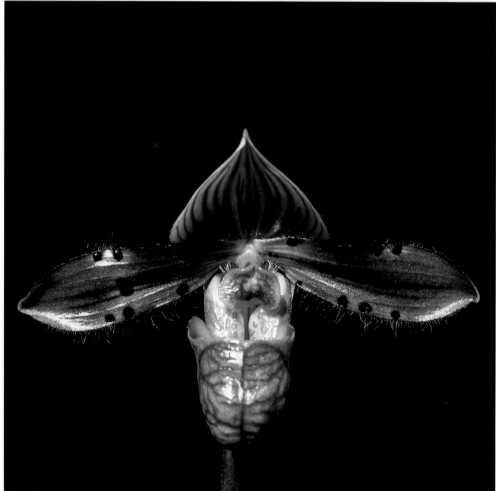

Right, Top: *Paph. purpuratum*

Right, Bottom: *Paph. venustum*

Opposite, Clockwise from Top Left: *Paph.* (*delenatii* x *rothschildianum*); *Paph.* (*rothschildianum* x Susan Booth); *Paph.* (*philippinense* x Maudiae)

Above and Opposite: *Habenaria lindleyana*

Subfamily: *Orchidioideae*

This is a huge family of highly successful terrestrial orchids that include the Mediterranean genus *Ophrys* and the widespread *Habenaria*.

Genus: *Habenaria*

With 500 species distributed throughout the tropics and in some temperate grasslands, *Habenaria* is one of the largest genera of terrestrial orchids. It is deciduous. After flowering the plants die down, leaving new, fleshy tubers on the rocky surface or underground. When the rainy season starts,

new green shoots appear. The flowers are borne terminally. In the prairies of North America, a purple *Habenaria* grows so extensively that the grassy expanse takes on a purple hue when it flowers.

The individual flowers, usually about 15 of them, are clustered at the terminal third of an erect stem. From the base of the lip, a long, thin, hollow spur extends backwards and a drop of nectar collects at its tip. The flowers are pollinated by moths and butterflies whose unrolled tongues can reach deep into the nectaries.

At least six species in the Thai-Malaysian region are showy and well worth cultivating — *H. carnea*, *H. dentate*, *H. lindleyana*, *H. medioflexa*, *H. rhodocheila* and *H. susannae*. The plants need careful handling and very little watering during the resting period. The main flowering season is September to October with some flowering in August and others extending to November.

Habenaria carnea

This lovely, delicate species is found on limestone in the lowlands of Pulau Langkawi, Kedah and Peninsular Thailand. It flowers in September. The pale pink flowers, about 4 cm long and 3 cm across, of round form, are arranged rather closely on an erect rachis about 20 cm tall. A superb specimen freshly collected from Kedah was awarded 'Best of Show' at the Singapore Orchid Show in 1962 and featured on the cover of the *Malayan Orchid Review*.

Habenaria lindleyana

This orchid is widely distributed throughout Thailand, flowering from August to November. It is one of the 13 species found in Thailand. The upright inflorescence carries up to 20 pure white flowers, each a maximum of 2.5 cm across. The slim spur is pale green and 4 cm long.

Habenaria medioflexa

This is an interesting Thai species with 30 greenish flowers and a white lip that is fringed like a grass skirt. The flowers are 2 cm across and appear from September to October.

Habenaria rhodocheila

This beautiful red-flowered species is distributed from northern Malaya to southern China and is abundant in Thailand, where it is commonly offered at the flower market for sale during the flowering season from September to November. In Thailand, many colour forms exist: various tints of pink, cinnabar to bright red, purple, and yellow to orange. A superb yellow variety grows on large granite boulders on Penang Hill. It is sometimes called the Aeroplane Orchid because of the shape of the lip which is 2.5 cm wide and 3 cm long.

Habenaria susannae

This striking white species grew in scrubland and paddy fields in Kedah, but it is now quite scarce in Malaysia. However, it has a wide distribution and is not in danger of extinction. It is also found in Hong Kong SAR. The plant grows to a height of 1–1.3 m and bears 5–7 pure white, fragrant flowers. The large striking lip is fringed with teeth along the edges of the sidelobes.

Subfamily: *Neottioideae*

The subfamily *Neottidioideae* is a large group of primitive orchids which are characterised by soft, granular pollen and a single anther whose cap falls off when the pollen is removed. They are terrestrial, herbaceous orchids with elongated, succulent or woody stems.

The beautiful Jewel Orchids, which are cultivated for their lovely foliage, are a popular tribe within this family. The strange mountain genus *Corybas* (the Helmet Orchids), the equally curious *Nervilia,* which produces inflorescences and leaves at different times of their growth cycle, and the saprophytic orchids constitute the second tribe, which is distributed in the Southeast Asian region.

The Jewel Orchids (Genera: *Anoectochilus, Goodyera, Haemaria* and *Macodes*)

Anyone who has ever looked at flowers has, sometime or other, been captivated by the beauty of the orchid flower, but only a diehard orchid lover would admire orchids for the appearance of the plant or its leaves. There are four exceptions to this, however, from the Malaysian-Indonesian region. We have four ancient genera of beautiful orchids whose leaves were described as appearing "as though they were woven from shimmering silken threads — a blending of cinnamon, purple, olive and moss green." These are the Jewel Orchids, the *Anoectochilus, Haemaria, Macodes* and *Goodyera,* which are also known by other names. They are all terrestrial orchids. *Anoectochilus* grows in the moist humus of the forest floor while *Haemaria discolor* and *Macodes petola* are found on rocks beside streams. *Haemaria discolor* is the easiest to cultivate and requires very little light, about the same amount as African violets.

The leaves of *Haemaria discolor* are purplish with red or gold veins and the surface is covered

by a metallic sheen. There is also a variety with green leaves which are marked with red. The white flowers are borne on erect spikes and last a week. The plant has succulent stems which ramble over the surface of the growing medium and, according to Dr. Yeoh Bok Choon, these stems are collected by Chinese herbalists who dispense them as a remedy for a whole range of chest ailments, ranging from asthma to tuberculosis. *Haemaria discolor* is quite an easy plant to grow. It can be planted in broken brick and charcoal or, alternatively, in burnt earth. It must be sheltered from rain and direct sunlight and can be placed together with the *Phalaenopsis* and *Paphiopedilum*. A French grower who was able to grow *Haemaria* into superb specimen plants recommended a potting mixture consisting of one-third leaf mould, one-third crushed peat and one-third sand, or 50 percent sphagnum moss and 50 percent tree fern. The plant grows very slowly and produces only three to five leaves a year. When the plant is young, flowering should be suppressed.

Macodes petola has heart-shaped leaves which are much larger and even more beautiful than those of *Haemaria discolor*. The flowers are similar to those of *Haemaria* but the lip is on top. *Anoectochilus* has leaves which are emerald green, veined with gold and silver. An interesting Javanese legend links the brilliant foliage of *Macodes petola* to the shreds of a magic silk scarf which had been placed on jagged rock by a beautiful goddess.

Opposite: *Macodes petola* is easily recognised by its iridescent foliage.

Top, Left: *Haemaria discolor*

Above: Inflorescence of *Macodes petola* in bud.

The Saprophytic Orchids and Their Allies

Genus: *Cryptostylis*

Cryptostylis are terrestrial orchids allied to the saprophytes, with some 10 species distributed from Sri Lanka to Australia. The few Malaysian species are not saprophytic. The name ('hidden style' in Greek) refers to the manner in which the column is enclosed by a small chamber formed by the lower part of the lip. The flower is pollinated by pseudo-copulation with male ichneumon wasps.

Cryptostylis arachnitis is found at 500–700 m, growing in dense shade, in Malaysia and the Philippines.

Genus: *Nervilia*

Members of this genus have underground tubers which flower before they produce leaves. The inflorescence is erect, with one to six flowers, with narrow petals, usually in a shade of green. After the inflorescence has disappeared, the tuber sends out a single heart-shaped plicate leaf which is deciduous. After a suitable resting period the cycle is repeated.

Subfamily *Epidendroideae*

The members of the enormous subfamily are extremely varied but they are all sympodial orchids with waxy pollinia. Several genera have succulent pseudobulbs which serve as storage organs. Most are epiphytes and some are terrestrial.

Left: *Arundina graminifolia*

Opposite: *Acriopsis javanica*

The popular *Dendrobium*, *Cattleya*, *Cymbidium* and *Odontoglossum* are well-known tribes in this subfamily. These orchids are extremely easy to grow because of the pseudobulbs and tough rhizomes.

The *Arundina* Tribe

This is a small tribe with four genera, *Claderia*, *Thunia*, *Arundina* and *Dilochia*. Only *Arundina* is cultivated.

At first glance, the flowers of *Arundina* may be mistaken for *Cattleya*, except that they are rather papery and the lateral sepals are close together, almost hidden by the lip. But the *Arundina* is a terrestrial orchid from the Old World, with a wide distribution that extends from Sri Lanka to northern India, southern China, Hong Kong SAR, Indochina, Malaysia, Indonesia and the Caroline Islands as far as Tahiti. (*Cattleya* is native in South America.) Introduced into the Hawaiian islands, *Arundina* has now become naturalised and may be found growing wild together with *Spathoglottis* and a few other terrestrial orchids.

Holttum has examined different varieties of *Arundina* collected throughout Peninsular Malaysia, and he concluded that they all belong to one species, *Arundina graminifolia*, although there is a whole range of minor differences in colour, size and the proportions of the floral parts.

Arundina graminifolia

The pseudobulbs of *A. graminifolia* grow close together, bearing many grass-like leaves with overlapping sheaths. The inflorescence is held well above the plant and carries a succession of flowers at its tip, one or two opening at a time and each lasting three days. The flowers are large and striking, with white petals and sepals and a contrasting purple lip. A patch of yellow is present in the throat. In Indonesia, a white variety has been collected.

Arundina graminifolia is widely distributed throughout Peninsular Malaysia and is found in open, sunny places, never in the shade of the forests. At the hill stations, it is often found by the roadside and on ground which has been cleared. Sometimes it grows in rocky places, by streams. Its striking flowers are easily spotted, but in the lowlands it is not as abundant or floriferous as to produce a great display of colour. Apparently, in the Himalayas, its flowering is seasonal and spectacular.

Cultivation of *Arundina*

The wide distribution of the species is an indication of the ease of cultivation. Before orchid hybrids became plentiful in the 1950s, *Arundina* was widely cultivated in this region, and along the sides of older streets in Singapore it is still possible to find flowering pots of *Arundina* standing on pedestals in small gardens at the entrance of terrace houses or on top of brick walls. They grow much better if grounded and exposed to full sunlight and should be grown in quantity to compensate for the few flowers on the stem. The beds should have good drainage with broken brick at the bottom and good topsoil, well mixed with compost. The plants must not be planted too deeply. When newly planted, the stems can be supported by stakes. They grow well if given organic fertiliser. Old flower stems produce side shoots at the base of the inflorescence, and when these offshoots have produced a firm swelling at the base, they may be removed and planted into sand or gravel, where they will quickly produce roots. If cultivated in pots under glass, *Arundina* should be brought out during the warm months and given full sunlight to facilitate flowering.

The *Acriopsis* Tribe

This small tribe has two genera, consisting of a few species with small flowers that are merely of botanical interest.

This is a common lowland orchid of the Malaysian region. The flowers are small and delicate and resemble a tiny insect with outstretched wings. This appearance is due to the horizontal, narrow petals; the lower sepals are fused, and together with the tube formed by the lip and column, it mimics the body of an insect.

Acriopsis javanica

This orchid is widely distributed in the lowlands of Myanmar, Malaysia, Singapore, Indonesia and Papua New Guinea, growing in a wide range of habitats — commonly on forest trees, rubber trees and coconut palms in rather open places. It is a lovely botanical to have in the garden as it is free-flowering and

the inflorescence is long and attractive. It may be tied to a tree fern slab or straight onto the branch of a tree.

The *Bulbophyllum* Tribe

This is the largest genus in the orchid family comprising some 2,000 species which are widely distributed throughout the tropics. In Papua New Guinea alone, there are some 600 species — a fascinating challenge for the 'splitters'. They are supremely successful epiphytes invariably found in large colonies. The pseudobulbs carry a single fleshy leaf at its apex and the flowers arise from the base of the pseudobulb. Some species have leaves a metre long while others are minute. Many species have complex flowers of remarkable form. However, the genus has been avoided by growers until fairly recently because most *Bulbophyllum* of small vegetative proportion have pale, creamy flowers that do not match up to *Vanda*, *Phalaenopsis* or *Dendrobium*, while some of the *Bulbophyllum* with showy flowers are huge plants that require a lot of growing space — and in a cool house environment.

Only a handful of hybrids have been made. If you wish to grow a *Bulbophyllum*, make sure that it is not one which produces an unpleasant odour when it blooms. Singapore's National Orchid Garden is one of the best places to view *Bulbophyllum*.

Bulbophyllum blumei

This is a common lowland species widely distributed in Sumatra, Malaysia, Kalimantan and central Philippines. The single, slender, brick-red flower is 6 cm tall.

Left: *B. echinolabium*

Opposite, Left: *B. dearei*

Opposite, Right: *B. lobii*

Bulbophyllum dearei

This montane species found in Borneo, Johore and southern Philippines, at 700–1,200 m, requires a cool-house environment. The fragrant yellow flowers, 4–5 cm in height, are similar to those of *B. lobii* except that they can be recognised by the backward-pointing sidelobes on their highly mobile lips.

Bulbophyllum echinolabium

Peter O'Byrne says that this is the world's largest-flowered *Bulbophyllum* species, its flowers sometimes reaching up to 35 cm in length. Native to the lower montane forests of Sulawesi in Indonesia, it is now endangered due to over-collecting for sale in the United States and Europe. The inflorescence extends slowly and bears up to 10 flowers but the blooms open singly, each flower lasting about a week. The next flower appears two months later.

Above: *B. blumei*

Opposite, Left: *Cirrhopetalum* Elizabeth Ann AM/AOS

Opposite, Right: *B. lilacinum*

Bulbophyllum gusodorfi **var.** *johorense*

This species, albeit attractive, is not floriferous like *B. vaginatum*. The flowers have prominent pink bracts arranged like a lady's fan, 6–8 on an inflorescence. It is found in Sumatra and in the southern Peninsular Malaysian state of Johore.

Bulbophyllum lilacinum

This is a free-flowering lowland *Bulbophyllum* that produces multiple, short, chilly-shaped inflorescences densely crowded with numerous greyish-cream coloured flowers. The posterior surface of the petals and sepals are spotted and striped with brown, while the prominent lip is a dull yellow.

Bulbophyllum lobii

This is a striking montane species growing at 1,000–1,500 m in the mountains of Thailand, Malaysia, Borneo and the Philippines. The plant has stout rhizomes with strong pseudobulbs 3 cm in height and 2.5 cm in diameter, carrying leaves that are 20 cm long and 5 cm wide. The flowers are waxy, widely open, fragrant and of an orange-yellow coloration densely spotted with crimson over the sepals and striped with crimson along the petals.

Bulbophyllum macrobulbum

This is a beautiful white *Bulbophyllum* which is set off against the pink leaves and pseudobulbs of the plant.

Bulbophyllum medusae

This has the same distribution but is less common than *B. vaginatum*. There are many similarities in their vegetative structure and in the shape of their flowers.

Bulbophyllum vaginatum

This is a common roadside orchid in Singapore, forming massive clumps on the branches of old Angsana trees (*Pterocarpus indicus*) and rain trees (*Enterolobium saman*). It flowers gregariously and a single tree trunk may be covered by thousands of their pale yellow flowers. It is very common throughout Sumatra, Borneo and Malaysia.

Genus: *Cirrhopetalum*

Some taxonomists list *B. medusae* and *B. vaginatum* under a separate genus designated by John Lindley in 1824 as *Cirrhopetalum*. It is similarly designated by the Royal Horticultural Society's International Committee on Orchid Registration. Phillip Cribb explained that if the name was derived from the Greek *kirrhos*, it meant orange-tawny or pale yellow, referring to the usual colour of the flowers; but it was more likely to have been derived from the Latin *cirrus*, meaning 'tendril' or 'fringe', in reference to the shape of the petals and their clustered arrangement as a fringe around the tight circlet of flowers. The second explanation is very apt for these two species. However, many species bear only one (such as in *B. blumei*) or two (in *B. biflorum*) flowers on the scape.

The 30 species of *Cirrhopetalum* are spread from India across Southeast Asia to Papua New Guinea, with a concentration in Malaysia. Some species flower gregariously several days after a large fall in ambient temperature occasioned by heavy rainfall. Kiat Tan (Dr. Tan Wee Kiat) recommends that *Cirr. medusae* be subjected to a temperature drop of 10 degrees Fahrenheit, and a heavy drenching to make it flower.

The first hybrid, *Cirr.* Louis Sander (*longissimum* x *ornatissimum*), was registered in 1936 but only received an AM/AOS for

var. Crown Point in 1984. *Cirr.* Elizabeth Ann (longissimum x rothschildianum), made by J.C. Chambers, is also awarded (AM/AOS). *Cirrhopetalum* is interfertile with *Bulbophyllum*, and with over a thousand species to work with, the hybridiser has limitless possibilities.

Bulbophyllum lepidum

This species and several other look-alikes, such as *B. acuminatum*, *B. makoyanum* and *B. pulchrum*, also belong to the genus (or section) *Cirrhopetalum*, though their sepals are lacking in the terminal fringe-like tendrils. *B. lepidum* is widely distributed over most of Southeast Asia but not beyond Borneo in the east. It is found in the lowlands and up to 900 m. This orchid is quite popular with the collectors because it is easy to grow into a specimen plant and does not take up much space. The leaf and pseudobulb measure up to 18 cm in length. It flowers throughout the year with a peak during the rainy season from September to February. The flowers are arranged in a semi-circle on the tip of the rachis, like a fan, and are quite distinct. The sepals are white to cream, flushed with red to maroon on their proximal half. The petals, lip, anther and column are yellow.

The *Bromheadia* Tribe

The *Bromheadia* tribe is made up of several genera that are located principally in tropical Asia and Africa. The dominant genus in Malaysia is *Bromheadia*. *Polystachia*, another genus, has over a hundred members in tropical Africa and scattered representatives in Central America, but it has only one representative in Malaysia. A third genus, *Ansellia*, is represented by a single variable species, *Ansellia africana*, which is widely distributed in tropical and South Africa.

Opposite: *B. (Cirrhopetalum) medusae*

Above: A montane *Cirrhopetalum* flowering in the Cool House of the Singapore Botanic Gardens

Right: *B. (Cirrhopetalum)* Lion King

Genus: *Bromheadia*

Bromheadia are terrestrial or epiphytic orchids with long, slender stems bearing numerous leaves and a terminal inflorescence which is strongly characterised by regularly alternating bracts. The flowers (rarely many) open one or two at a time in two rows. The blooms are well displayed, with petals spread apart, resulting in a flat flower.

There are 11 species altogether, distributed from Sumatra and Malaysia to Papua New Guinea. Seven of these species occur in Peninsular Malaysia. Only *Bromheadia* is common to all areas.

Bromheadia finlaysoniana

The common terrestrial Singapore-Malaysian orchid is found in open country and developing belukar. It also occurs in Sumatra and Borneo. It was described and named by Lindley as *Bromheadia palustris* when it flowered in cultivation in the south of Wales in 1841.

Left: *Cirr. lepidum*

Below: *Cirr. picturatum*

This is an easy plant to grow, either in pots or bedded on the ground in an equal mixture of topsoil and garden mould. It enjoys organic fertiliser. The plants flower when they are a metre tall, with the larger plants producing bigger flowers.

The beautiful, white, fragrant flowers are striking in the early morning sunlight, held singly or (rarely) in pairs on tall stalks. Unfortunately, they wither by noon, and for this reason alone, the *Bromheadia* has not been commonly cultivated. The flowers take about three weeks to develop and they may be slowed down by dry weather. Normally, in a bush, one or two spikes carry a bloom, but after a rainy spell, full flowering of the entire clump occurs.

Left: *Bromheadia finlaysoniana.* At one time fairly common in Singapore gardens, it has largely disappeared because it is no match for the modern orchids, such as the new splash *Cattleya*, shown on the facing page.

Opposite: *Blc.* Haw Yuan Beauty 'Hong' AM/AOS

The *Cattleya* Tribe

At one time or another all orchid growers have been attracted to the *Cattleya* because of its large, delicate, colourful flowers which are among the most flamboyant in the orchid world. The name *Cattleya* refers to the most popular genus in a large tribe which comprises *Brassavola, Laelia, Cattleya, Sophronitis, Epidendrum, Diacrium, Schomburgkia, Broughtonia, Cattleyopsis, Laeliopsis, Leptotes* and a few newly designated genera such as *Barkeria, Rhyncholaelia, Sophronitella* and *Neocognitanxia*.

The *Cattleya* tribe is native to Central and South America, but its species and hybrids are now widely grown all over the world. As they are popular with Asian hobbyists and nurserymen, they are briefly covered in this volume.

Cattleya is commonly bred with *Laelia* because of the great similarity and complimentary elements between the two genera: better form in *Cattleya* and better colouring in *Laelia*. *Cattleya* has also been bred with the montane *Sophronitis* to produce red *Cattleya*, and with *Brassavola* to enlarge the lip and add frills to it. The multi-generic hybrids referred to as *Laeliocattleya* (*Lc.*), *Sophrolaeliocattleya* (*Slc.*), *Brassolaeliocattleya* (*Blc.*), *Potinara*, etc., have many features in common, and for convenience many growers just refer to them by the collective name 'Catts', or *Cattleya*, growing them together and treating them in the same way. In fact, they have slightly different temperature requirements, depending on parentage, and they ought to be separated into cool-growing types (with *Sophronitis* or some other montane species featuring prominently), intermediate types and warm-growing types. Each needs a different night temperature to flower well.

Cattleya and its relatives thrive in the temperate and subtropical regions, such as California, Florida, Taiwan and even France. The genus does reasonably well more than 15 degrees away from the equator. It tends to become vegetative when it is grown in the lowlands along the equatorial belt, but selected species and hybrids flower fairly freely in Singapore. Most species have a definite flowering season, and hybrids have been made between species which flower at different times of the year in order to extend the flowering season. Unfortunately, *Cattleya* are slow-growing, each pseudobulb flowering only once, and therefore they can never be as free-flowering as the evergreen *Dendrobium, Oncidium* and the monopodial orchids. Thai breeders are resorting to miniatures and crossing with *Epidendrum* and *Broughtonia* to develop plants that could flower twice or, hopefully, several times a year.

The flower form is most perfect when the flowers open slowly, and they will do this when the ambient temperature is low: a glance at the list of awards from the American Orchid Society will show that most *Cattleya* awards are made in late autumn, winter and early spring.

Cattleya are sympodial orchids which are most easily grown in clay pots. Practically every medium has been used for growing these plants, which like to be left on the dry side. Watering frequency depends on the locality and the medium used for potting. If they are grown in the open in the tropics, it is recommended that the amount of nitrogen be severely curtailed during the rainy season.

Left: *Lc.* (Mildred Rives x *C.* Yaigwavong)

Genus: *Cattleya*

Cattleya are tough, sympodial orchids which may be subdivided into two groups on the basis of their foliage. The unifoliate *Cattleya*, which are sometimes referred to as the *labiata* group, have a single large leaf per pseudobulb, and they were the first *Cattleya* to be introduced into Europe. They all have large flowers which are accentuated by a large lip. *C. labiata* is the best known species in this group and it embraces perhaps a dozen previously designated species which are now regarded as varietal forms of *C. labiata*. The bifoliate *Cattleya* have two (and occasionally three) leaves per pseudobulb. The flowers are produced in clusters of five to as many as 20 per stem. Old Singaporean growers are familiar with the delicate, soft textured *C. bowringiana*, but many of the bifoliate *Cattleya* are in fact thick-textured and waxy, and they come in a vast array of colours and colour patterns. Some species, such as *C. guttata*, *C. walkeriana* and *C. granulosa*, are singularly striking and have impressed modern hybridisers with their potential.

Genus: *Laelia*

Laelia are closely related to *Cattleya*, differing from the latter in having eight pollinia while *Cattleya* have four (cf. *Dendrobium* and *Eria*). The flowers are smaller, with narrower sepals and petals which are brilliantly coloured — in yellow, orange, bronze, tan and deep violet. Some species, such as *Laelia flava*, have a long inflorescence. This genus comprises some 30 species which are epiphytic or lithophytic in Mexico, Central America and Brazil. Hybrids with the lowland species do very well in Singapore but those from the highlands do not.

However, several *Laeliocattleya* will grow and flower in Singapore, and a number of these are bicoloured. The popular *Laeliocattleya*

that are suited to cultivation in the tropical lowlands are *Lc*. Derna Anderson, *Lc*. Amber Glow and *Lc*. Dorset Gold.

Genus: *Sophronitis*

A small genus of dwarf plants confined to the mountainous regions of Brazil, *Sophronitis* is responsible for the scarlet colouring of such famous hybrids as *Slc*. Falcon 'Alexanderi', *Slc*. Alexander 'Westonbrit' and *Potinara* Gordon Siu. The last-mentioned hybrid flowers occasionally in Singapore.

Sophronitis grandiflora is a jewel, but all *Sophronitis* need to be grown in a cool house. Recently, Singapore's Woon Leng Orchids made *Slc*. Red Zac which is free-flowering, deep red and miniaturised.

Genus: *Brassavola*

This genus, with some 20 species distributed over Central and South America, are *Cattleya*-like plants with terete or almost terete leaves. The flowers have narrow sepals and petals but an enormous lip. In *Brassavola digbyana*, the large lip extends up to 10 cm across, ruffled along the edge and fringed with fine hair-like appendages: the flowers are greenish-white and have a citrus scent. It will flower in Singapore.

Genus: *Epidendrum*

Epidendrum is the largest genus of orchids in the New World with over 1,000 species, many of which occur in great abundance in

On these two pages: **Three 'Catts' that do extremely well from Singapore to Bangkok: the white** *Lc*. **Hawaiian Wedding Song** (opposite, top); **the pink** *Blc*. **Pink Diamond** (opposite, bottom) **and the bicoloured** *Blc*. **Alma Kee FCC/OST** (left).

Central America. Its name is derived from two Greek words, *epi* and *dendron,* which when combined means 'upon a tree', referring to its growing habit. During the 18th century, many epiphytic orchids from the Old World were referred to as *Epidendrum,* even by Carl Linnaeus.

Epidendrum are sympodial and divided into two groups on the basis of their vegetative characteristics:

1. Reed-stemmed *Epidendrum,* which have no swollen pseudobulbs
2. Pseudobulbous *Epidendrum*

Above: This is an unusual splash petal bifoliate *Cattleya,* with a beautiful crinkled edge on its purple-flushed petals. It is fragrant in the morning.

Right: *Bc.* (Country Road x Orchid Library) is a large, well-shaped white, unifoliate *Cattleya* that will flower only once a year, around February to March. It is useful as a stud plant, serving to improve the size and shape of the more free-flowering bifoliate *Cattleyas* that are more suited to tropical lowland cultivation.

Reed-stemmed *Epidendrum* can be grown into bushes on the ground, and several varieties can withstand full sun. Their flowers are produced in clusters on a long inflorescence and come in various shades of orange, yellow, white, light green and tan. They can be incorporated into landscapes as they are quite easy to grow and require little care. *Epidendrum radicans* may be seen in some commercial establishments in Singapore.

The pseudobulbous *Epidendrum* are compact plants which can be grown together with the evergreen *Dendrobium* and *Ascocenda*. *Epidendrum atropurpureum* and *Epidendrum tampanse* grow quite well in Singapore. Some hybrids have been raised with *Cattleya*, and a few have found their way into Southeast Asia. *Epicattleya* have delicate, round flowers with a prominent lip and are very attractive.

Above: *Blc.* Chai Lin

Opposite, Clockwise from Top Left: *C.* Chyung Guu Swan; *Pot.* Hwa Yuen Gold; *Blc.* Sanyung Ruby; *Blc.* Chunyeah var. Tzeng Wen AM/AOS

Although *Epidendrum* is an enormous genus, few *Epidendrum* species are represented in orchid collections in Southeast Asia. To a large extent, this is due to the ban on the import of plants from tropical and South America into the Asean region because of the vital role of rubber in the national economies. Any orchid grower from this region who is interested in *Epidendrum* should look at the *Epicattleya* from Hawaii, which have the best features of both *Epidendrum* and *Cattleya* and are really more rewarding to grow.

These small, bifoliate *Cattleyas* represent the effort being put in by breeders in Thailand and Singapore to produce plants that will bloom readily and several times a year in a warm climate. They put an emphasis on yellow and red, these being two contemporary colours sought after by growers in Southeast Asia.

On these two pages, Clockwise from Top Left:
Blc. (*briegeri* x Waikiki Gold); *Potinara* Free Spirit
'Lea' AM/AOS; *Slc.* Red Zac; *Slc.* Rimfire;
Blc. Haadyai Delight; *Lc.* Tutti (lemon strain);
Lc. Tutti (orange strain)

Genus: *Diacrium*

This is a small genus, with about six species distributed in Guatemala, Venezuela and Trinidad. It is closely allied to *Epidendrum*. The plants are epiphytic or lithophytic, with hollow pseudobulbs bearing tough, leathery leaves. The inflorescence is upright, tall with many flowers. *Diacrium bicornutum*, the Virgin Orchid, is a graceful species with pure white flowers. It has been bred with *Cattleya*, producing some very fine hybrids, such as *Diacattleya* Chastity, which are suited to tropical lowland cultivation.

Genus: *Broughtonia*

This small genus from the West Indies is only known because of *Broughtonia sanguinea*, a warm-growing, dwarf species which flowers throughout the year. The arching rachis carries round crimson flowers, 1 cm long, in a small terminal cluster. Some hybrids have been created.

Genus: *Schomburgkia*

This Mexican genus has large pseudobulbs which serve as a nesting place for ants and they are troublesome to collect in the wild. The inflorescence is tall, 1–2 m long and carries many *Laelia*-like flowers only towards the tip, facing all directions. The petals and sepals are much twisted, and the pedicle is long, an altogether clumsy appearance for a flower. *Schomburgkia* will flower in Singapore but, being more of a botanic curiosity, are not common in private collections.

Opposite, Top: *Bc.* Mem. Vida Lee

Opposite, Middle: *Bc.* Maikai

Opposite, Bottom Left: *Bc.* Hot Spice

Opposite, Bottom Right: *Epidendrum* Hokulea

Left, Top: *Epicattleya* Roman Holiday

Left, Middle: *Bc.* Binosa

Left: *Cattleytonia* Why Not

Above: *Epidendrum* (Star Valley x Joseph Lii)

The *Coelogyne* Tribe

This is a fairly large group of orchids which is made up of four genera, *Dendrochilum*, *Pholidota*, *Chelonistele* and *Coelogyne*, with over 300 species, all native to the tropical and subtropical regions of the Old World. Only the genus *Coelogyne* is of horticultural interest.

Members of the *Coelogyne* tribe are characterised by having:

1. A single-jointed pseudobulb which bears one or two leaves and a terminal inflorescence;
2. A hooded or winged column; and
3. A three-lobed lip which is concave or saccate at the base and marked by keels and crests.

Genus: *Coelogyne*

Within the tribe, the genus *Coelogyne* has all the species with large, showy flowers. This is a large genus with 150 species distributed from the Himalayas and South China across Southeast Asia to the New Hebrides. The better known species come from the lowlands, but some of the smaller montane species are more colourful. The genus is divided into five sections, with four sections having flowers which open simultaneously and one where the flowers open one at a time. At one time, several lowland species of *Coelogyne* were often cultivated, but because the flowers tend to be short-lived, they have not been able to compete with the ubiquitous long-lasting *Dendrobium* hybrids that have come to capture the attention of the orchidists. *Coelogyne* can be grown in pots and given the same treatment as *Cattleya* and *Oncidium*, and in hanging containers, their pendulous sprays will be shown to best advantage. As they will grow into large clumps, some growers put them out on specially raised beds of charcoal bordered by coral, or into an orchid landscape among other epiphytic orchids. They like good drainage, some humus (such as tree fern or redwood bark) and strong organic fertiliser. The larger plants will always flower better than small ones. Some species are gregarious in flowering.

Coelogyne asperta

A common species widely distributed from sea level to 500 m in Southeast Asia from Malaysia to Papua New Guinea, this plant is easy to cultivate, and flowers freely. The inflorescence is 30 cm long and many flowered. The flowers are 6 cm across, creamy white and pleasantly scented. The lip is a contrasting brown and covered with warty processes over the midlobe. Unfortunately, the attractive flowers last only five days on the plant, and even fewer if they are removed.

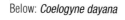

Below: *Coelogyne dayana*

Right: *Coelogyne asperta*

Coelogyne foestermani

This is a beautiful species with long sprays of up to 15 large, fragrant, white flowers made even more spectacular by the gregarious flowering which occurs about three months after dry weather. A lowland species, it grows abundantly on high trees in Sumatra, Malaysia and Kalimantan.

The pseudobulbs are narrow, 12 cm long, with horizontal grooves, and the dark green leaves are 30 cm by 4.5 cm. The inflorescence is up to 40 cm long, arched, arising as a separate shoot at the base of a vegetative pseudobulb (*heteranthous*) with 8–15 floppy flowers. The sepals and petals are 3–4 cm long and 1 cm wide.

Coelogyne longifolia

This species belongs to the section *Longifoliae*, which is characterised by a long, gradually lengthening inflorescence which bears only one or two flowers at a time. Jim Comber, an authority on the orchids of Indonesia, wrote in the *Malayan Orchid Review* that

Opposite, Left: *Coelogyne rochussenii*

Opposite, Right: *Coelogyne pandurata* (Photo by courtesy of John Teoh)

Above: *Coelogyne longifolia*

the inflorescence may attain a length of 80 cm. The apple green flowers are only 4 cm across with narrow sepals and petals. New flowers continue to appear for several months. This species is found in East Java, most commonly at 1,600 m.

Coelogyne pandurata

This delightful, light green *Coelogyne* is a hardy lowland species growing on old trees near rivers in Sumatra, Malaysia and Kalimantan. The flat pseudobulbs are 12 cm long, 7 cm wide, 2.5 cm thick, rather well spaced with leaves which are 50 cm long. It grows strongly if properly cultivated. The inflorescence is 30 cm long with 12–15 flowers of pale green; the lip is white with black markings.

Coelogyne rochussenii

This is one of the commonest riverine lowland species in Peninsular Malaysia. It is also found in Indonesia and the Philippines. Generally not free-flowering, it sometimes produces a spectacular display after a period of aseasonal weather. Each of Mak Chin On's dozen plants of *C. rochussenii* once carried 50–60 flower spikes with 2,000–2,500 flowers in Singapore.

The grooved pseudobulbs are 20 cm long and bear large leaves 30 by 10 cm that are wide at the top. The slender, pendulous, 50 cm inflorescences arise from the base of lead pseudobulbs and carry many lemon yellow flowers with narrow parts. The inner surface of the sidelobes of its white lip is marked with brown.

Right: *Chelonistele perakensis*

Opposite, Left: *Dendrochilum saccolabium* — plant (top) and inflorescence (bottom)

Opposite, Right: *Pholidota parviflora*

Genus: *Dendrochilum*

The *Dendrochilum* is a rich, relatively unexploited group of fascinating orchids which will delight any grower of botanicals. From Myanmar to Papua New Guinea, with concentrations in the mountains of Sumatra, Borneo and the Philippines, most are found at high altitudes (1,000–3,000 m) and several species are confined to a single valley. With small pseudobulbs and sculptured leaves, *Dendrochilums* are easily grown into specimen plants carrying multiple inflorescences, provided one has a cool house. Many members have long arching sprays with numerous flowers arranged in two rows. They radiate a beautiful crystalline iridescence under a magnifying lens.

No breeding has been undertaken for this genus which is probably interfertile with *Coelogyne* and *Pholidota*, to which it is related. Beautiful specimen plants have been shown in the United States, and some attempt has been made to grow these plants in Singapore.

Genus: *Pholidota*

This is a small genus with 40 species which share many similarities with *Dendrochilum* in the appearance of the plants and flowers and growing conditions.

Genus: *Chelonistele*

Described by Holttum, this genus has the habit of *Coelogyne* and a flower form intermediate between *Coelogyne* and *Pholidota*. The single Malaysian species is *Chelonistele perakensis*, which is similar to the Phillipine *Pholidota* (*Chelonanthera*) *ventricosa*.

The *Cymbidium* Tribe

The *Cymbidium* is perhaps the first orchid to be cultivated, and today its numerous hybrids are popular as house plants and also as cut flowers. It is one of the two genera that make up the *Cymbidium* tribe. The other genus *Grammatophyllum* is distinguished by having two members that are the largest orchids in the world, in terms of both plant and flowers size. We refer to the Tiger Orchids, *Grammatophyllum speciosum* and its close relative *Grammatophyllum papuanum*. A plant of *Gramm. speciosum* collected from Penang in the 19th century for the orchid firm of Frederick Sander weighed a tonne. Divisions of the plant that were given to the Singapore Botanic Gardens still bloom at the Gardens, where they never cease to amaze visitors. Its inflorescence is 2 m long, bearing 40 flowers 10 cm in diameter, but these are dwarfed by those of *Gramm. papuanum*, also 2 m long with 40 flowers, but a whopping 25–30 cm across!

Members of the *Cymbidium* tribe are terrestrial or epiphytic orchids that enjoy a wide distribution from India, China and Japan to Australia and the Solomon Islands in the east and Madagascar in the west. Their leaves are long and slender, closely arranged on stems and jointed at the base. The inflorescence is lateral, long, erect and arching, at times pendulous with few or numerous flowers. The sepals and petals are similar in appearance, and the lip is trilobed with longitudinal ridges. Very few *Cymbidium* species flower in the tropical lowlands, but it is an outstanding species for orchidists living in the temperate regions. On the other hand, several species of *Grammatophyllum* flower readily at sea level near the equator. Holttum has long advocated the hybridisation of *Cymbidium* with *Grammatophyllum,* but this has not been carried out with any noticeable success. However, breeders in Southeast Asia have recently shown interest in making primary hybrids within *Grammatophyllum.*

Genus: *Cymbidium*

Cymbidium is still probably the most popular spray orchid in temperate countries. These cool-house types are developed from the handsome montane species native to the foothills of the Himalayas and the mountains of Myanmar and Yunnan. The tropical lowland *Cymbidium* are generally not spectacular, and only a single hybrid has been registered by Singapore. *Cym. chloranthum*, which flowers in the lowlands, is outstanding, but one has yet to see its hybrids on the market. In China, growers admire the leaves and fragrance of the *Cymbidium*, paying scant attention to their flowers. Chinese growers appear to stick with the local species that have small flowers and slim, dark green leaves.

Cymbidium are either epiphytic or terrestrial plants possessing short pseudobulbs and long narrow leaves. Fifty species are distributed from Madagascar through India, China and Japan to Australia. The inflorescence arises from the base of mature pseudobulbs that will flower only once. The flowers of heavy substance lasts for several weeks on the plant, and many species have large flowers. The lip is large and commonly marked with coloured dots or splashes, and the midlobe has two longitudinal keels. They are easily grown in pots in a mixture of burnt earth and charcoal, and they will accept a variety of potting mixes. *Cymbidium* grow well in the tropics, but high night temperatures will cause the buds to abort. Night temperatures below 18 degrees Celsius are absolutely essential when the plants are in bloom. Perhaps the problem could be solved by crossing miniature *Cymbidium* with the lowland species, but no one has been willing to undertake the challenge.

Cymbidium chloranthum

This elegant *Cymbidium* is a small plant with thin leaves up to 45 cm long by 3.5 cm wide. The inflorescence is erect, 45 cm long with 20 pale green flowers that become a light crimson when the pollinia are removed. This species is found in Sumatra, Java, Borneo and Peninsular Malaysia.

Cymbidium ensifolium

This miniature, terrestrial *Cymbidium* is probably the oldest cultivated orchid species, being the favourite *lan fa* alluded to by Confucius (circa 6th century B.C.). It is distributed from Japan to China southwards to Indonesia, occurring at 700 m above sea level in tropical areas. It flowers readily, though not abundantly, in Singapore. It gives off a delicate fragrance in the morning, and it is often cultivated in glazed pots so that the plant can be brought indoors to be admired when it is in bloom.

Cymbidium finlaysonianum

This is the commonest *Cymbidium* of the lowlands, being widely distributed throughout the coastal areas of Sumatra, Malaysia and the Philippines. It can still be seen flowering on roadside trees in Singapore. It grows in exposed locations, in large clumps, especially in Penang and the northern half of the Malay Peninsula.

The narrow leaves are thick and fleshy, up to 75 m long and 4 cm wide. The long, pendulous inflorescence may reach a length of 90 cm. Flowers are spaced rather far apart and are of a dull, olive colour, but they are accentuated by

Opposite: *Cym.* Blooming Alexander

Overleaf: *Cym. finlaysonianum*

a dark purple lip which is splashed with yellow. The tepals are of very heavy substance. The erect petals point forwards. The plant flowers from June to November, the flowers usually lasting a fortnight.

Genus: *Grammatophyllum*

Grammatophyllum is a small genus with between six and 12 species. The reason behind this range in the number of species is because some taxonomists distinguish species with subtle differences in the shape of the lip and include *Grammatophyllum stapeliiflorum* (syn. *Grammangis stapeliiflora*). They are distributed from Myanmar in the west and eastwards through Indochina, the Malayan archipelago, the Philippines, Papua New Guinea and up to the Solomon Islands. These are all large

Opposite, Clockwise from Top Left: *Cym.* Black Magic; *Cym. lowianum* is a common garden plant in Kunming and Lijiang which are 2,300–2,500 m above sea level in Yunnan province, China; *Cym.* Napolian Glitter; *Cym.* Jung Fran. These cool-growing *Cymbidium* hybrids have large flowers that are accentuated by a contrasting lip. The flowers are of heavy substance and last many weeks even when they are harvested. Such *Cymbidium* are therefore very appealing, and they are often imported during Christmas and the New Year. Unfortunately, they require vernalisation (very low temperatures for a few months) if they are to bloom again. Hence, they are not suitable for tropical lowland culture.

Right: *Gramm. scriptum* var. Green Envy ABM/OSSEA (1980) illustrates the similarities between *Cymbidium* and *Grammatophyllum*. In Singapore, Dudley Leicester grew this clone into a large specimen plant that produced a dozen sprays up to a metre long, with thousands of flowers.

plants with long arching sprays that form a delightful focal point in a medium-sized or large garden, and they should be grown more extensively. Several species have made a regular appearance at orchid shows in the region in the past decade.

Grammatophyllum may be divided into two sections on account of the differences in the shape of the pseudobulbs:

1. *Pattonia*, with tremendously elongated, cane-shaped, cylindrical and sometimes pendulous pseudobulbs, as characterised by *Gramm. speciosum*;

2. *Gabertia*, with ovoid pseudobulbs, as characterised by *Gramm. scriptum, Gramm. martae, Gramm. measuresianum*, etc.

Section: *Pattonia*

Grammatophyllum speciosum

This is an enormous plant, the largest orchid in the world. A single plant can weigh a tonne. Its fleshy cane-like stems are up to 3 m long and carry many leaves which are up to 60 cm in length and 3 cm in width. It flowers only once a year: around July in Penang, July to October in Thailand and either in January or in July in Singapore. However, the plants may not flower consistently and may alter their flowering season. Nevertheless, the plant is a great centrepiece because the numerous sprays are magnificent and the flowers do last a long time on the plant.

The inflorescence is 2 m long with 40 flowers, each 10 cm across. The flowers are dimorphic, with abnormal forms restricted to the lower flowers which commonly possess only two petals and two sepals, and are without a lip or column. They are set far apart, and the upper flowers enjoy an excellent spacing and arrangement. The petals and

sepals are a pale greenish yellow, overlaid with chocolate blotches.

Gramm. speciosum is found throughout the lowlands of Southeast Asia to Papua New Guinea. In cultivation, it does best when it is treated as an epiphyte in a forest setting, that is, when tied to the crotch of a large tree. There has been some attempt to return this species to the forest in Singapore, but we are still waiting to see the results. It can also be grown in large containers. When established, it requires full sunlight. One must be careful in selecting a location for this orchid because the plant will eventually become far too large to move. The last time the Sultan of Johore exhibited *Gramm. speciosum* at the Singapore Flower Show, the services of a crane and a large truck were required to move the plant.

Grammatophyllum papuanum

Its flowers are similar to *Gramm. speciosum*, but the lip is different, and its flowers are larger. Holttum gave it a distinct species rating. Andree Millar illustrated an excellent clone of *Gramm. papuanum* with well rounded flowers in *Orchids of Papua New Guinea*.

Section: *Gabertia*

Grammatophyllum scriptum

Perhaps the most popular cultivated species because of its compactness and well arranged blooms, *Grammatophyllum scriptum* exists in a number of forms. The albino variety, which has green flowers, is particularly attractive.

Grammatophyllum scriptum is found in lowland forests from Sulawesi eastwards to Papua New Guinea and some Pacific Islands. Andree Millar reported that it was very common on coconut trees in old plantations in Papua New Guinea. The pseudobulbs are up to 20 cm long and 9 cm wide, clustered into

Grammatophyllum speciosum

Top: The abnormal first flower on the inflorescence has two petals and two sepals and is devoid of its crucial reproductive parts.

Above: The normal flower

Opposite: *Gramm. scriptum*

large clumps of 20 or more pseudobulbs. The white roots are erect and dense and completely surround the pseudobulbs. The arching inflorescence is up to 2 m long with up to 30 blooms that are 8 cm across. The common colour is yellow to pale green marked by brown bars. They last for several weeks on the plant. The flowering period is December to March.

The two pure-coloured varieties, yellow and green (albino), are devoid of bars.

It may be tied to a tree in a sunny location or grown in a basket. It does not like frequent repotting, and some plants can be lost when repotting is attempted.

Grammatophyllum martae

In vegetative appearance, this is similar to *Gramm. scriptum*, but it occurs in the lowlands and low foothills of the central Philippines. The inflorescence is 1–1.5 m long and carries up to 80 well formed flowers, 7.5 cm across, with broad petals and sepals. Unfortunately, the flowers, a dull brown with a thin greenish edging, are not eye-catching. It would be interesting to breed *Gramm. martae* to other species of *Grammatophyllum*.

Grammatophyllum measuresianum

A lowland orchid endemic to the Philippines, *Gramm. measuresianum* has longish pseudobulbs measuring 25 cm by 6 cm. An inflorescence of 1.5 m carries up to 80 flowers that are 5 cm across. They are densely marked with chocolate brown spots and blotches over a pale green to creamy base. The three lobes of the lip are striped on both their medial and lateral aspects.

Grammatophyllum multiflorum var. tigrinum

A compact plant with pseudobulbs 15 cm in length and 4 cm in diameter, *Gramm. multiflorum* var. *tigrinum* produces long arching sprays of up to 1.5 m carrying a hundred blooms 5 cm in diameter which are bright yellow with brown blotches. The species is variable and some have an olive green base; in some the markings are not so clear. Holttum includes this species under *Gramm. scriptum*.

Left: *Gramm. martae*

Overleaf, Left: *Gramm. measuresianum*

Overleaf, Right: *Gramm. multiflorum* var. *tigrinum*

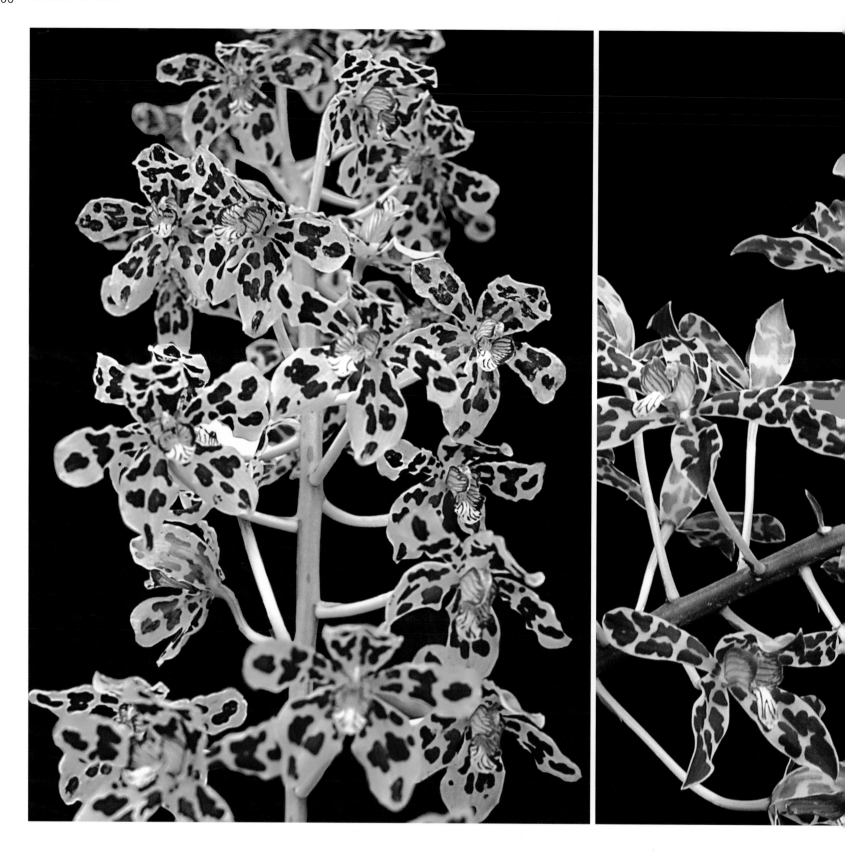

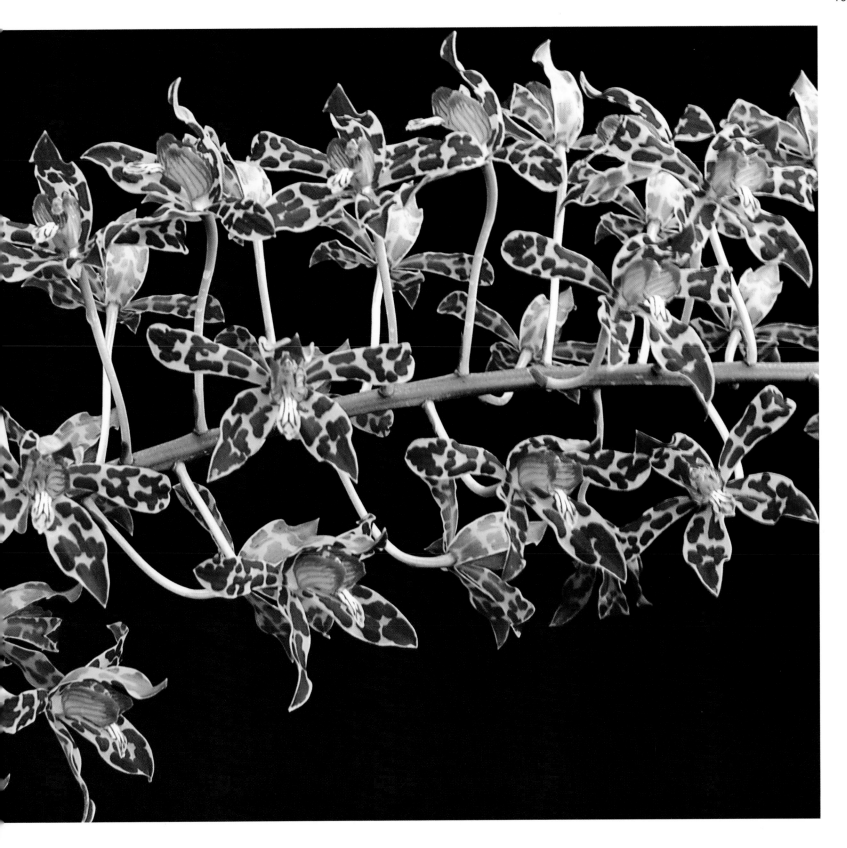

The *Dendrobium* Tribe

This is one of the largest tribes in the orchid family, comprising two principal genera, *Dendrobium* with 900 species and *Eria* with 400 species. A third member, *Porpax*, is numerically a small genus. Papua New Guinea appears to be the centre of development of *Dendrobium*.

Flickingeria was split off from *Dendrobium* in 1965. It consists of some 50 species. The members of the *Dendrobium* tribe are spread throughout Southeast Asia extending to South China, Japan, India, Sri Lanka, Australia and New Zealand. They are easy to recognise because of obvious similarity in the flowers which arise from the axils of cane-shaped pseudobulbs. The inflorescence is lateral (arising from leaf axils), with single or numerous flowers arising from a small group of bracts. The column extends to the column foot, joined on both sides by the base of the lateral sepals to form a mentum. The sepals are triangular, and the petals are different from the sepals, either smaller or larger. The lip has three lobes and a narrow base which is connected to the column foot to form a spur. *Dendrobium* has four pollinia; *Eria* has eight.

When one mentions *Dendrobium*, most of us immediately form a mental image of either the *Phalaenopsis*-type or horn-type *Dendrobium*. But, in fact, the type plant belongs to an entirely different section. It is *Dendrobium moniliforme*, a plant already described in detail in the Chinese *Herbals* written in the 8th century. *Dendrobium moniliforme* and other deciduous *Dendrobium* belonging to the section *Eugenanthe* (Peter O'Byrne refers to this as the section *Dendrobium*) were collected as medicinal plants over a thousand years ago. The collection proceeds unabated because of demand.

Genus: *Dendrobium*

The ever popular *Dendrobium* is one of the most rewarding orchids to grow, being handsome, floriferous and hardy. It is the second largest genus with its more than 900 species. The genus has a wide distribution extending from Korea and Japan through the Indo-Malaysian region to Papua New Guinea, Australia and the Pacific Islands, with a concentration of extremely fine species in Papua New Guinea and a different pocket in Thailand and Myanmar.

Dendrobium is derived from two Greek words, *dendron* and *bios* which mean 'tree' and 'life' respectively. The name refers to the epiphytic nature of the orchid — living on a tree. *Dendrobium* are pseudobulbous orchids of diverse vegetative structure and floral form. Their fleshy pseudobulbs may vary from a few centimetres in some species to a few metres in others. The leaves are commonly flattened, although in some species they are terete and jointed at the base. Some species are deciduous. The inflorescence is usually terminal or subterminal, with few or many flowers. The lateral sepals are usually triangular and joined to the end of the column foot to form a mentum. The base of the lip is also joined to the column foot, sometimes spreading to the side so as to form a spur. The large genus *Dendrobium* has been divided into 20 sections. From the horticultural viewpoint, there are two major subdivisions: the warm-growing *Dendrobium* which need daily watering, and the cool-growing deciduous *Dendrobium* which require a cool dry season to induce flowering. Hybridisation in the tropical lowlands (mainly in Singapore, Malaysia, Thailand and Hawaii) has concentrated on only two sections, *Phalaenanthe* and *Ceratobium*, while in the temperate countries,

Right: *Den. thyrsiflorum*

horticultural interest is focused on the members of *Callista* and *Eugenanthe,* which belong to the deciduous group. Over the past two decades, there has been some interest in the intersectional breeding between the *Latourea* and *Nigrohirsute* sections and the popular *Phalaenanthe* and *Ceratobium* sections, following the introduction of colchicine-induced, fertile amphidiploid *Dendrobium* hybrids by Professor H. Kamemoto of Hawaii.

Section: *Phalaenanthe*

Dendrobium bigibbum

This famous Cooktown Orchid occurs in northern Queensland and in Papua New Guinea. It has the appearance of a dwarf *Den. phalaenopsis* with thinner leaves and pseudobulbs up to 45 cm long. The flowers are rounder, rosy purple or white. For registration purposes, *Den. bigibbum* is regarded as a variety of *Den. phalaenopsis,* although hybridisers continue to designate their breeding stock as *Den. bigibbum. Den.* Macrobig is an interesting, valuable, amphidiploid intersectional hybrid between *Den. macrophyllum* and *Den. bigibbum.*

Dendrobium phalaenopsis

This is one of the finest orchids for the beginner (and for the expert), being easy to cultivate, free-flowering, and always exceptionally beautiful. It is native to the Tanimbar Islands of Indonesia (Timor Laut) and grows well throughout the tropical lowlands, adapting even to the subtropical regions without requiring special housing.

The pseudobulbs are fleshy, 30–60 cm long, swelling from a slender base and bearing stiff, pointed leaves at the top of the stem. The inflorescence is terminal, sometimes two or three appearing together on a strong

pseudobulb. It is up to 40 cm long and carries 6–15 flowers, 6–8 cm across. The flowers are well arranged in two rows on either side of the stem. In good clones, the petals are rounded and touch or overlap. A characteristic feature is the double mentum formed by the lip and the column foot: this is best appreciated when the flower is viewed from the side. The best colour forms are believed to be deep magenta or pure white, but there is now considerable interest in flowers which have a lavender blush at the centre fading into white at the periphery. The paler and more delicate colour forms are preferred by the cut-flower industry. Peak flowering is from August to November and the sprays may last up to three weeks on the plant.

Dendrobium phalaenopsis was introduced into England in 1855, and soon several outstanding clones were recognised and given varietal names — *Schroderianum, Hololeucum* and *Rothschildianum,* the latter two referring to white varieties. At least three varieties

which found their way into Hawaii were tetraploid (Extra, Giganteum, Ruby), and they were the forerunners of extremely outstanding, polyploidy, *Phalaenopsis*-type hybrids such as *Den.* Diamond Head Beauty, *Den.* Lady Hamilton, *Den.* Lady Fay, *Den.* Lady Hay, *Den.* Anouk and *Den.* Hickam Deb. *Den.* Pompadour, though less outstanding than these hybrids in flower form, outclassed them in vase life and it was extensively propagated for cut flowers in Thailand, and at one time also in Malaysia and Singapore.

Hawaii, Singapore and Thailand produced a wide array of hybrids by crossing *Den. phalaenopsis* with *Ceratobium.* These so-called intermediate *Dendrobium* form the bulk of the *Dendrobium* flower export trade from these countries. More recently, intersectional hybrids have involved crossings between *Den. phalaenopsis* and members of the sections *Latourea* and *Nigrohirsute.*

Dendrobium superbiens

Den. superbiens is a variable species with the same distribution as *Den. bigibbum,* and it is probably an intersectional hybrid between the *Phalaenanthe* and *Ceratobium* sections. The inflorescence is arched, 40 cm long, carrying 10–15 flowers 5 cm across, of rich violet purple to rose purple. The petals are narrow at the base, fanning out distally, and are rounded at the tips. The sepals are broad, curved backwards and have a wavy outline. The lip is short and has the double mentum of a *Phalaenanthe.* It grows well in Singapore and was extremely popular up to the late 1950s. Unlike *Den. phalaenopsis,* it flowers throughout the year. *Den. goldiei* is another name for this 'species'.

Section: *Ceratobium* (syn. *Spatulata*)

The members of this section are known as the horn or antelope *Dendrobium*, a term that alludes to the shape of their dorsal petals. They are also called cane *Dendrobium* after the shape of their pseudobulbs.

At the World Orchid Conference in Miami in March 1984, Phillip Cribb pointed out that John Lindley had established the section *Spatulata* with six species of *Dendrobium* in 1843. Notwithstanding the fact that it was also John Lindley who proposed the section *Ceratobium* in 1850, according to the rules endorsed by the International Code of Botanical Nomenclature established in 1978, the former name takes precedence, and *Ceratobium* must be relegated to the status of a synonym for the section *Spatulata*. I have kept to *Ceratobium* by force of habit.

Dendrobium antennatum (syn. *Den. d'albertsii*)

A dwarf *Ceratobium*, this species has short, slender pseudobulbs 15–20 cm tall bearing light green leaves. The inflorescence is 15–35 cm long with up to 10 well spaced, upright flowers 7 cm high. The narrow, pointed petals are erect, light green and slightly twisted, while the sepals are white and curled backwards. The white lip has five keels and is marked with violet. A common plant in the coastal areas of Papua New Guinea and its adjacent islands, it flowers readily in Singapore.

Opposite: *Den. phalaenopsis* var. *alba*

Right: *Den. antennatum*

Dendrobium canaliculatum

Among the antelope *Dendrobium*, this species is so unique that it almost merits a separate class. The plants are dwarfed (4–16 cm tall), forming clusters of short pseudobulbs which resemble onions. The short, strapped leaves accentuate the similarity. Several inflorescences arise near the tip of the pseudobulbs. The spike is 20–25 cm long and carries 10–20 flowers, 2.5 cm across, towards its distal third. The flower form is rather open. The petals and sepals have a half twist; white towards the centre, the flower segments become suffused with yellow to green from the point of the twist distally. The lip is purple.

Den. canaliculatum is distributed from southeastern Papua New Guinea to the coastal regions of northeastern Queensland. An Australian variety, *Den. canaliculatum* var. *nigrescens*, is orange.

The colour, form, shape, floriferousness and compactness of *Den. canaliculatum* are transmitted to its hybrids. They make excellent houseplants in the tropics.

Dendrobium discolor

The commoner, alternative name for this species, *Den. undulatum*, is certainly more descriptive of this orchid, whose unmistakable flowers have twisted sepals and petals with wavy, undulating edges. The sepals are also twisted backwards. *Den. discolor* is distributed throughout Queensland and Papua New Guinea, being the commonest *Dendrobium* in the lowland areas of the latter.

The pseudobulbs are tall, up to 120 cm, with leaves of 15 cm length on its upper half. It is easily grown into a specimen plant which will produce a massive display of flowers. The inflorescences are produced terminally and at leaf axils, each 45 cm long, with 30 flowers closely arranged all round the rachis.

Den. discolor is an extremely variable species, in the colour and the size of its flowers and in its vegetative form. The common variety is a dull yellow with fine purple stippling on the petals and sepals. The variety *Den. broomfeldii* is a clear yellow; the variety *d'albertsii*, which Millar refers to as the 'Rigo Twist', is miniature and the flowers are of a beautiful bronze colour. The different varieties flower at different times of the year, and their flowers last for three months.

Den. discolor was one of the first orchids to be used for hybridising in Singapore. In 1933, John Laycock crossed it with *Den. lasianthera*, another Papua New Guinean cane *Dendrobium*, to produce *Den.* Constance, which served as a forerunner of other fine cane *Dendrobium,* such as *Den.* Brown Curls, *Den.* Curlylocks and *Den.* Parkstance. Many primary hybrids were later produced from *Den. discolor,* such as *Den.* Dang Toi (x *goldiei*), *Den.* Caprice (x *Den. macrophyllum*), *Den.* Ursala (x *veratrifolium*), *Den.* Medusa (x *violacea-flavens*), *Den.* Champagne (x *mirbelianum*) and *Den.* Pauline (x *phalaenopsis*). Some fine yellow and yellowish-green cane *Dendrobium* have been bred from *Den.* Champagne and *Den.* Ursula.

Dendrobium lasianthera

This is a very tall cane *Dendrobium* which may grow to a height of 3 m. Its inflorescence is 40 cm long with 10–20 flowers closely arranged along the distal half of the erect rachis, a point for its use in hybridising cut flowers. The flowers are 6.5 cm tall, with the petals much twisted and narrow, the sepals twisting once and curving backwards with folded edges. The common colour is a lacquered brown. The lip is large and marked with purple.

Native to Papua New Guinea, this species has been thoroughly studied by Andree Millar, who suspects that the finer forms are probably

natural hybrids between *Den. lasianthera* and *Den. discolor.* Due to their great height, hybrids of *Den. lasianthera* are more suitable for landscaping in large gardens than for inclusion in the collection of small hobbyists.

Dendrobium lineale (syn. *Den. gouldii,* Den. veratrifolium)

This floriferous species was one of the first orchids to be introduced into Singapore, via Indonesia. It was a tremendous success from the start because of the inspiring appearance of a well grown plant. At the 2nd Singapore Orchid Show in 1932, a plant of *Den. veratrifolium* shown by the famous John Laycock carried 77 inflorescences and well over a thousand flowers. This free-flowering characteristic is happily transmitted to its primary hybrids, and crosses such as *Den.* Louisae (x *Den. phalaenopsis*) and *Den.* Caesar (x *stratoites*) were among the first *Dendrobium* to appear on the cut-flower market.

The slim wiry canes of *Den. lineale* may grow to a height of 1.5–2 m, but they are

Opposite: *Den. discolor*

Above: *Den.* Curlylocks. The cross was made by the Singapore Botanic Gardens. Its seed was sown in November 1941 and flowering occurred in 1949. It thrives in full sun.

generally 60 cm to 1 m tall, with dark green leaves on the distal half of the stems. They are easily grown into large specimen plants, either in pots or attached to tree stumps. Several inflorescences are produced at the leaf axils on a single cane, each 75 cm long, with 30 or more flowers. There is a wide range of colour forms, but the petals are commonly white to cream and the lip is white-splashed and streaked with purple. The dorsal sepal is curved backwards while the petals are upright and twisted, from a half twist to three full turns. Andree Millar thought that the variations in shape and colour of the blooms are both due to the existence of natural hybrids, with back crosses in all directions between *Den. lineale* on the one hand, and *Den. mirbelianum* and *Den. warianum* on the other.

The white forms occur in the lowlands of Papua New Guinea and on the islands of New Britain, New Ireland and Bougainville. They grow by the thousands on the giant *Calophyllum inophyllum* and on old coconut palms.

Dendrobium mirbelianum

This is a dwarf *Ceratobium* from the Moluccas and Papua New Guinea, with canes 20–60 cm long. The inflorescence is 30 cm long with 12–15 flowers 5 cm across and of good form. The sepals and petals are a light yellow-green spotted with violet brown, and the lip has five keels, all green and veined with purplish brown. This is an attractive species, but unfortunately it is cleistogamous, that is, self-pollinating; the flowers fade within two or three days of opening and each sets seed. Nevertheless, by selfing the plant, the Singapore Botanic Gardens produced plants with long sprays carrying 30–40 flowers that last for two months.

In its native habitat, *Den. mirbelianum* is an extremely variable species, with 11 synonyms. Self-pollination probably contributed to the maintenance of individual strains.

Short canes and a 30 cm inflorescence with 30–40 flowers are very desirable characteristics of a possible stud plant. A number of free-flowering *Dendrobium* have been produced in Singapore by crossing with *Den. mirbelianum*. Two famous Singapore Botanic Gardens hybrids are *Den.* Noor Aishah (25 percent *Den. mirbelianum*), named after the wife of Singapore's first president Yusof Ishak, himself an avid orchid grower, and *Den.* Tien Suharto (12.5 percent *Den. mirbelianum*).

Dendrobium schulleri

A favourite among hybridisers, *Den. schulleri* is a common species from northern Papua New Guinea. A clone with uniformly green flowers

features in the parentage of the numerous, hugely successful hybrids from Hawaii and Singapore. In its native Papua New Guinea, *Den. schulleri* is found in the company of *Den. lasianthera* and *Den. smillieae*, epiphytic on the exposed branches of trees growing near water, at swamp edges, along creeks and lagoons.

The stems are 100 cm long, about average

for a *Ceratobium*, with dark green leaves 17 by 7 cm. The plants start flowering when they are still quite small. The inflorescence is 30–50 cm long, horizontal and arching, with up to 30 flowers, 6 cm across, of a pale brownish green. The lip is greenish white with purple markings. The petals are slightly twisted. It has some resemblance to

Den. mirbelianum, but it is much finer and possesses broader petals. Some forms have clear lemon yellow flowers.

A plant of *Den. schulleri* var. Cher which received a Cultural Commendation Certificate in the United States was described as "6 feet in diameter, with 84 canes, 43 in flower with one to seven inflorescences each. The flowers were too numerous to count."

Among the long and distinguished list of green and bicoloured hybrids which have been made with *Den. schulleri* are *Den.* May Neal (x Hawaii), *Den.* Kwa Siew Tee (x Morgenster), *Den.* Noor Aishah (capra x Champagne), *Den.* Mary Trowse (x Hula Girl), *Den.* Pikul (x Indonesia) and *Den.* Yong Kok Wah (x Mary Trowse). *Den.* May Neal is a fine parent giving rise to numerous second-generation green and yellow hybrids, such as Gold Flush, Liholiho, Siam and Sri Siam.

Dendrobium stratoites

Do the flowers remind you of soldiers with fixed bayonets? That was the reason given by H.G. Reichenbach in 1880 when he chose the name for the species. The species occurs in the lowlands of Papua New Guinea and the Moluccas in large clustered clumps with long spindle-shaped stems up to 2 m in length and 2 cm in diameter, carrying leathery, apple green leaves, 8 cm long and 2 cm wide, getting progressively smaller towards the tip of the cane. The leaves stay for several years on the plant. On a rachis of 15–20 cm length, there are 5–15 flowers which are well displayed; stiffly erect because of its good substance; large (8–9 cm tall and 3 cm across); with creamy

Opposite: *Den.* Tien Suharto

Right, Top: *Den.* Alice Spalding 'David' HCC/OSSEA (1967)

Right, Bottom: *Den. stratoites*

white sepals, thin, twisted greenish sepals and a showy pointed lip that is marked with violet stripes. They are long-lasting and the plant blooms throughout the year.

This is a typical hardy cane *Dendrobium*. It thrives and flowers well in Singapore and was used to make some of the early great hybrids, like *Den.* Caesar (x *phalaenopsis* var. *schroederianum*), an extremely floriferous plant admired for its readiness to grow into a specimen plant.

Dendrobium taurinum

A handsome though uncommon species from the Philippines, this has canes which are 120 cm tall. Up to 21 flowers are borne on an erect 50 cm spike. The 6 cm tall, two-toned flowers are distinguished by a huge lip which is white in the centre, keeled and bordered with deep purple at its tip and edges. The white, cream or greenish sepals are rolled backwards and waxy at the edges. The broad, deep purple petals are erect and are twice twisted.

The floral shape of *Den. taurinum*, in particular the bull-lip, dominates its hybrids.

Section: *Latourea*

This section has rather interesting greenish flowers which are accentuated by a large, contrasting lip. It is distributed in Papua New Guinea and the surrounding islands and comprises some 35 species. Only three have received the attention of growers and hybridisers — *Den. atroviolaceum*, *Den. macrophyllum* and *Den. spectabile*. The last species has the dubious honour of being possibly the one orchid which can be described as grotesque. Members of this section are interfertile with members of the sections *Phalaenanthe*, *Ceratobium* and *Nigrohirsute*.

Dendrobium macrophyllum

From the early days of orchid cultivation, this has been a common pot plant in Java. It is

On these two pages, from Left to Right: *Den.* Benezir Bhutto; *Den. taurinum*; *Den. tangerine*.

distributed from Papua New Guinea to Java at a level of 600–2,000 m. It prefers light shade and a somewhat cooler environment than what we can offer in Singapore. The flattened pseudobulbs, which are 20–30 cm long, bear inflorescences of 30 cm length, carrying up to 15 flowers, 4–4.5 cm across, of yellow to pale green. They are hairy on the outside. The lip is a pale green marked with purple veins on the sidelobes. The flowers are very unusual but not really attractive.

Section: *Callista*

This is a small section with species restricted to Myanmar, southern Thailand and Peninsular Malaysia. The erect pseudobulbs bear few leaves. The inflorescence is born near the apex and carry many flowers that are distinguished by broad petals and sepals and a round lip devoid of sidelobes. The lip is generally of an intense yellow colour and hairy within, and it encloses a short nectary. All the member species are beautiful.

Dendrobium farmeri

In the opinion of the late Professor Eric Holttum, the former authority on Malayan orchids, this is perhaps the most beautiful native species. Unfortunately, its flowers last only a week on the plant.

Distributed from Myanmar down the isthmus to northeastern Peninsular Malaysia up to northern Pahang, *Dendrobium farmeri* is a beautiful species which has enjoyed long horticultural patronage. The first plants that reached England were collected in Malaya and sent to WF Farmer of Cheam in Surrey, hence the name. Plants collected from Malaysia flower freely in Singapore but those from Myanmar do not. In its natural habitat, *Den. farmeri* grows on large trees overhanging rivers, and the plant is usually shaded throughout

the year. It prefers light shade and dampness during the growing phase and does well as a hanging plant.

Dendrobium thyrsiflorum

Native to northeastern Thailand, its flowering season is February to April. The magnificent, lantern-shaped inflorescence carries 30–50 creamy white flowers that are offset by a bright yellow lip which intensifies into a deep orange at the throat. Each flower is 3–4 cm across and gives off a delicate fragrance. Unfortunately, they only last 5–7 days on the plant.

This species grows well but flowers poorly under normal warm conditions in Singapore. The 45 cm long cylindrical pseudobulbs are marked by vertical grooves and bear 4–6 pairs of dark green leaves near the apex.

Section: *Eugenanthe* (syn. *Dendrobium*)

The name of this section was proposed by Rudolf Schlechter to designate the true *Dendrobium* (*Eu-dendrobium*) which is distributed in Myanmar and Thailand, with an extension northwards to China and Japan, and southwards to Malaysia, Indonesia, and the Philippines, eastwards to Papua New Guinea and westwards to Sri Lanka. The flowers are borne on bare pseudobulbs which are erect or pendulous. They appear on many nodes simultaneously. Thus, a flowering plant is always striking even though the individual inflorescences are short and few flowered. The flowers are large, many are extremely attractive and there is a good range of colours within this section.

Through excellent hybridisation programmes and masterly use of tetraploidy, breeders in Hawaii and Japan have produced an excellent range of hybrids within this section. Such hybrids and species that are

Top: *Den. macrophyllum* (*Den. musciferum*)

Above: *Den. thyrsiflorum*

Opposite: *Den. fameri*

about to flower will readily do so when they are imported into Singapore, and their flowers last a long time. Unfortunately, while the plants do well vegetatively thereafter, they will not produce such spectacular flowering again in the tropical lowlands.

Dendrobium anosmum

A lowland orchid which requires a regular dry season, this species does extremely well in Penang where it is a common sight in many homes, but it is hard to cultivate in Singapore. It thrives when tied to a living branch which allows unrestricted growth of its pendulous stems which may grow to a length of 120 cm or more. The flowers are 6–10 cm across, in a mauve purple. The lip is covered with fine hair and marked by two deep purple blotches. There is a single flower per node, but when the plant is well grown, every node bears a flower. The flowering season is February to April. The flowers are fragrant. Peter O'Byrne likens the scent to that of raspberry jam. The white variety is very attractive.

Dendrobium cretaceum

This white *Dendrobium* is native to Assam in India and Tenasserim in Myanmar, but it is not commonly planted because it does not flower well in the tropics. The plant illustrated in Chapter 2 was photographed recently at the Mist House of the Singapore Botanic Gardens. Its flowers are 3 cm across and of a beautiful white.

Dendrobium fimbriatum

Once discovered on Gunung Pantai in the southern Malaysian state of Johore, this species is usually collected from Nepal and Myanmar.

It requires a cool dry season to flower well and is better suited to northern Peninsular Malaysia and countries further north, but it will occasionally flower in Singapore. The stems are 120 cm long, pendulous and coloured a light yellow green when they are old. Six or more beautiful golden yellow flowers, 5 cm across, are carried on pendulous inflorescences. The round lip is a deep orange and fringed at the edges; hence the species name.

Dendrobium fredericksianum

This species is endemic in the lowland forests of eastern and southeastern Thailand. Its blooming period is January to March. Each inflorescence carries 3–4 flowers, 5.5 cm across and of an attractive chrome yellow. The lip is of a slightly darker yellow, and some clones bear two small maroon blotches at the throat. The flowers are waxy, of good substance and last for up to five weeks on the plant.

Flowering in Singapore is not as bountiful as in its native Thailand.

Dendrobium pierardii

Native from northeastern India southwards to the northern border of Peninsular Malaysia, *Den. pierardii* are more likely to be imported at the resting stage from Myanmar or Thailand. The stems are slender, pendulous and 100 cm long. The flowers are produced along the lower half of the deciduous stems. They are 5 cm across, of a delicate mauve with a cream coloured or pale yellow lip marked with purple at the base. The lip is hairy inside, extends outwards and is rounded at the tip. There is a gentle gracefulness about the plant which does extremely well in the subtropical region. Unfortunately, it does not flower well a second time in Singapore.

Top: *Den. pierardii*

Above: *Den. fredericksianum*

Opposite, Clockwise from Top Left: *Den. fimbriatum*; a pink Yamamoto *Dendrobium* hybrid; *Den. anosmum* var. *alba*

Hybrids of *Eugenanthe*

Hybridisation in this section has been dominated by Jiro Yamamoto of Japan for over 30 years. Working with polyploid parents, he created a remarkable list of sensational hybrids famous throughout the world and acknowledged as being in a class of their own. They are popularly known as the Yamamoto *Dendrobiums* and are instantly recognisable.

Section: *Nigrohirsute* (syn. *Formosae*)

The section *Nigrohirsute* has some 35 species which are distributed from the Himalayas southwards to Southeast Asia. The name is derived from the blackish hair which covers the leaf sheaths. Many members of this section have handsome flowers which are long-lasting, but very few do well in the tropical lowlands.

Dendrobium cruentum

This species is found on small trees in open forests in the lowlands of Peninsular Thailand. It flowers well in Singapore. In Bangkok, it flowers continuously throughout the year. The pseudobulbs are slim, 30 cm long, with many leaves and 1–3 flowers on short inflorescences which appear at the upper half of the pseudobulb. The flowers are 4–6 cm across and pale green on the sepals and petals. The lip is marked by five bright red keels at the base. The sidelobes of the lip are also flushed with red. The flowers last about a month on the plant.

The few hybrids with *Den. cruentum* display a marked dominance of this parent.

On these two pages, Clockwise from Top Left: *Den. cruentum; Den. bracteosum* var. *alba; Den. dearei*

Dendrobium dearei

A lowland species with long-standing popularity, *Den. dearei* produces long-lasting white flowers throughout the year. It is native to the Philippines and northern Kalimantan. The pseudobulbs are 60–100 cm tall, and the flowers are 5 cm across. The short inflorescence carries up to a dozen flowers. The mature leaves and stems are devoid of black hair.

Dendrobium sanderae

This is a beautiful montane species from the Philippines which needs to be grown in a cool house. Flowering once a year, it produces multiple large inflorescences with up to 12 white flowers that are 5–6 cm wide, marked with purple at the throat. They last for about 14 days. Flowers of the variety major have flowers that are 10 cm across.

Section: *Pedilonum*

Seldom regarded as ornamental plants, some species in this section are actually quite stunning when they bloom in profusion, examples being Papua New Guinea's *Den. smillieae* and *Den. purpureum* and the Malaysian *Den. secundum*. These species are now receiving the attention they so richly deserve. *Pedilonum Dendrobium* have slender fleshy stems covered with well-spaced leaves, but inflorescences are borne only on leafless stems, with some species having few flowers and others bearing large clusters on a relatively short inflorescence. The lip is usually prominent, long and pointed.

Dendrobium bracteosum

For durability of blooms, few orchids can challenge *Den. bracteosum*, whose flowers last up to six months on the plant. They are produced in clusters of 5–8 flowers at several nodes on mature stems, the clusters coalescing to form a bounteous bouquet of white, cream or pink. The flowers are 3 cm across, with a lingulate, yellow lip and are fragrant. The main flowering season is winter to spring.

This is a compact plant with several cylindrical stems 20–40 cm long which become furrowed with age. The leaves are thin, narrow, pointed, tough and up to 9 cm long. Mature pseudobulbs are leafless. It is epiphytic in lowland forests and mangrove and is distributed throughout Papua New Guinea and New Ireland. A large, dark red form occurs on Rossel Island in Papua New Guinea.

It would be interesting to see whether its lasting qualities could be transmitted to its hybrids.

Dendrobium smillieae

Known as the Bottle-brush Orchid of Australia, this is a robust orchid found in open forest in the lowlands and foothills of Papua New Guinea and Queensland, Australia. It is met in a wide range of habitats as an epiphyte on swamp trees, old coconut trees and other jungle trees and even as a lithophyte on rock faces and rocky outcrops. It grows into large clumps with slender canes 15–100 cm long, 1–2 cm in diameter. Old canes are clothed in thin, membranous sheaths. The 4–15 cm long inflorescence is borne at the apex of each cane and packed on all sides, each flower only 1 cm across and 1.5–2 cm long. The petals and sepals spread out like a fan and are commonly a pale pink with a greenish flush at the tips. The lip is marked by a shiny green pouch.

Right: *Den. leonis*

Far Right: *Den. sanderae*

Opposite, Clockwise from Top Left: *Den. uniflorum; Den. smillieae; Den. purpureum; Den. secundum.*

A white variety has been collected from the Sepik Swamps of Papua New Guinea in the company of *Den. lasianthera* and *Den. coranthum*. In the ordinary variety, the canes turn brown when they mature, but in the white variety, they become yellow. At the 2003 Singapore Orchid Show, an *alba* form of *Den. smillieae* was awarded Best in Show.

Dendrobium purpureum

A native of Sulawesi and Papua New Guinea, *Den. purpureum* is an epiphyte of dense lowland forests up to a level of 800 m. The short, bottle-brush or pincushion inflorescence is borne at the nodes of deciduous canes. The flowers are a light purple, tipped with green.

Dendrobium secundum

This is a common orchid present in abundance throughout the lowlands in the eastern half of Thailand. It also occurs in Myanmar, Indochina, Malaysia, Indonesia and the Philippines and adapts readily to different cultural conditions. However, it likes bright sunlight and a period of rest before flowering. This point is important for the Singapore grower to remember because of the heavy rains that occur from November to January. In its native habitat, *Den. secundum* generally flowers from February to April. The short inflorescence (7.5–12.5 cm) is densely packed with pink or pure white flowers arranged in rows, all facing one direction. The flowers retain their conical form, opening just enough to reveal the contrasting orange lip. It is an interesting orchid to cultivate but unfortunately, the flowers are short-lived.

Section: *Aporum*

Dendrobium leonis

This is a common lowland epiphyte with creamy, indistinct flowers of a pleasant vanilla fragrance produced singly on the stem.

Section: *Distichophyllum*

Dendrobium revolutum (*uniflorum*)

This bushy *Dendrobium* is found in the lowlands of southern Malaysia, whereas the smaller-flowered *Den. uniflorum* is found near the peak of Mount Ophir in Johore and at the Taiping hills in Perak, Peninsular Malaysia. They are related species, and the flowers are very similar. The flowers are solitary, facing

downwards, with white sepals and petals that are pointed and curled backwards. The lip is a dull greenish yellow and is marked by three linear brown lines which form the keels along the midlobe. The midlobe is folded along its midline, and the two halves turn downwards laterally. Being a large plant, it is not sufficiently showy to be cultivated as a garden plant.

Dendrobium crumenatum

The common Pigeon Orchid is found on roadside trees all over Singapore and Malaysia, even in the heart of bustling cities. A Malay name for this orchid is *bunga angin*, meaning 'wind flower', referring to the impression that the plants are transported by the wind to the trunks of trees. It has a wide distribution extending from Southeast Asia to India and China. It grows in exposed locations in the lowlands and is the only native orchid which is abundant in Singapore.

The lower ends of the pseudobulbs are swollen and yellow but become furrowed with age and taper into slender stems up to a metre long, bearing thick, leathery leaves. The white flowers arise singly or in pairs from the bract-like sheaths along the distal part of the stem. They are 5 cm across and fragrant but last only a few hours. *Den. crumenatum* flowers gregariously 8–11 days after a sudden drop in temperature occasioned by tropical thunderstorms. The same plant flowers several times a year (see Chapter 11 on Control of Flowering).

Hybrids of *Dendrobium*

Hardy, colourful, floriferous, with handsome inflorescences that carry large flowers which last a long time in the vase, the horn and *phalaenopsis Dendrobium* and their intermediate hybrids fulfill every requirement for use as a tropical landscape plant and a commercial cut flower. The improvements that are constantly sought by hybridisers are an increase in colour range, density, colour clarity, striping, blush and dual tones; rounder forms, stiff upright flowers; increase in the length of the flower stem and in the number of flowers; symmetrical arrangement of flowers on the stem; retention or increase in the floriferousness of the plant and elimination of bud drop. Polyploid or amphidiploid parents are commonly employed to achieve these goals as well as to extend the vase life of the cut *Dendrobium*. Recently, premium on space has emphasised the importance of compactness of plant form, and many delightful hybrids have resulted through breeding with *Den. canaliculatum*.

Since most of the desirable qualities for tropical *Dendrobium* are present only in the sections *Phalaenanthe* and *Ceratobium*, hybridisation has been dominated by crossings between the two sections. The resultant intermediate *Dendrobium* form the backbone of the orchid export trade of Singapore, Malaysia, Thailand and Hawaii. These hybrids are too numerous to name, and every year new members join their rank. In this volume, only a few can be illustrated. The large number of species in *Ceratobium* and the vast variation in colour forms even within the species open up limitless possibilities for hybridisation within the genus.

In the 1970s, Dr. Clair Russell Ossia wrote an excellent review of the antelope *Dendrobium* for the *American Orchid Society Bulletin*. She counted 56 hybrids involving *Ceratobium* between 1900 and 1949. Between 1950 and 1954, another 15 were added, then nine more between 1955 and 1959. Between 1960 and 1964, 66 appeared, and the figure rose to 93 in the next five years. This coincided with the rise of the international orchid export trade and the hosting of a World Orchid Conference in Singapore. Unfortunately, the

Left: *Den.* Kwa Siew Tee

Opposite, Left: *Flickingeria comata*

Opposite, Right: *Den.* Nilda Patricia Zedillo

zeal for hybridisation within the genus has been dented by the advent of mericloning and the flooding of the market with high-quality *Dendrobium*. Never to be daunted, we offer here some good advice from Dr. Ossia.

Members of the subsection *Minacea* (*Den. antennatum, Den. lasianthera, Den. stratoites*) impart the following valuable traits to their hybrids: long-lasting flowers (up to six months in certain cases), floriferousness (numerous flowers and frequent blooming periods), compactness in plant habit and stiffly erect petals.

The subsection *Mirbeliana* (*Den. mirbelianum, Den. capra, Den. gouldii*) provides the yellows, greens and blues. When in bloom, the hybrids flower profusely and the flower spikes may carry 40 or more flowers. The flowers last well, cut well, and have more rounded form when bred with the *Phalaenopsis* group.

Hybrids of the *Taurina* species (*Den. taurinum*) inherit the striking large lip and bright colour. The subsection *Platypetala*

(*Den. schulleri, Den. violacea-flavens, Den. canaliculatum*) imparts a compact habit, sprays with many flowers and helps improve the shape and substance of the flowers. Hybrids of the subsection *Undulata* (*Den. discolor*) have tremendous substance and frilly flower shapes.

Novelty enjoys a separate niche, but one person's novelty is another's abomination. Much of the novelty arose from genetic aberrations produced through mericloning, and they are not the result of deliberate breeding.

Genus: *Flickingeria*

Formerly listed as section *Desmotrichum* within the genus *Dendrobium, Desmotrichum*

was assigned a separate genus status in 1961 by Alex Hawkes, who named it *Flickingeria* after an editor of an orchid journal. It has some 50–70 species which are common lowland orchids enjoying a wide distribution from Sumatra to Papua New Guinea. The ends of the branched stem each carry a single leaf. Flowers are produced singly. They are pale yellow marked with purple. Wavy, yellow hair radiate from the midlobe of the lip.

Flickingeria comata

This is perhaps the commonest lowland species growing on trees in estates and parkland. Although the plant is free-flowering and the lip is interesting, the blooms last for barely a day.

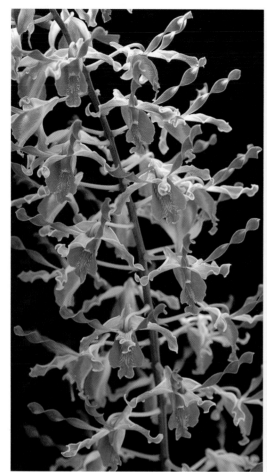

The VIP *Dendrobium* Orchids of Singapore. Famous visitors to Singapore often have their visit commemorated with the naming of an orchid hybrid of their choice. On these two pages, we feature some of the current hybrids.

Above: *Den.* Margaret Thatcher

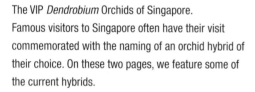

Right, Top to Bottom: *Den.* Masahiko and Hanako; *Den.* Elizabeth; *Den.* Siew May (wife of former Singapore President Ong Teng Cheong)

Opposite, Clockwise from Top Left: *Den.* Roh Moo Hyun Yangsuk; *Den.* Mother Teresa; *Den.* (Peter Furmiss x Elizabeth x Michiko) — this *Dendrobium* is waiting to be named; *Den.* Istana (tetraploid) — a promise of more good things to come

Genus: *Eria*

The genus *Eria* is closely allied to the *Dendrobium*, sharing many similarities in vegetative structure and in the form of the flowers, but *Eria* can be distinguished by the presence of eight pollinia, while *Dendrobium* has only four pollinia.

Eria is a large genus with 350 species widely distributed from Sri Lanka to China and India, and southwards across Southeast Asia to the Pacific Islands. Many species have handsome flowers and are very showy when in bloom. They are easy to handle, and many lowland species grow well and flower frequently in tropical gardens. Montane species are well suited for cultivation in temperate countries and are sufficiently hardy to be grown outdoors. They all require some shade and frequent watering. The simplest way is to tie the pseudobulbs to a slab of tree fern, although they are also well suited to pot cultivation.

No hybrids have yet been made with *Eria* despite the obvious potential for some very interesting interspecific crossings. They are probably compatible with some *Dendrobium*.

Eria albido-tomentosa

This species belongs to a section of *Eria* which is characterised by short, swollen pseudobulbs and a hairy inflorescence bearing numerous flowers. The flowers open a few at a time, with the apex continuing to grow for several weeks. The flowers are pale green with a dull purple lip and are hairy on the posterior surface. They are long-lasting. This species is distributed from Myanmar and Thailand through Malaysia to Sumatra and Java.

Eria atrovinosa

This is a spectacular species which was first collected from Pahang in Peninsular Malaysia. The plant is small, with narrow, thin, pointed, flowing leaves, 18 cm long (rather like a miniature *Cymbidium*), 2–3 leaves per pseudobulb. The inflorescence is erect, 15 cm long, bearing numerous, bicoloured flowers which are very striking. The bracts are cream while the rest of the flowers are a deep crimson. It is a lowland species and flowers well in Singapore.

Eria bractescens

Widespread from Nepal to Papua New Guinea, this is a common, moderate-size epiphyte of lowlands up to 1,000 m. The 7 cm pseudobulbs bear up to three 20 cm leaves and an erect 15 cm inflorescence with a dozen cream-coloured flowers 2 cm across. The lip is marked with pink on the keels and purple at the side. It is a very attractive species but is not particularly free-flowering in Singapore.

Eria braddonii

This species has the largest flowers among the *Hymenaria* group of Malaysian *Eria* and has prominent sulphur yellow bracts at the base

of each flower. The flowers are white with some red markings at the base of the sepals and petals, and the lip is yellow with patches of pink and red. It occurs quite commonly along river banks in the lowlands of Peninsular Malaysia and Borneo.

Eria cepiflora

A common lowland species in Malaysia and Kalimantan, this is also found in hills up to 1,400 m. The inflorescence is erect and densely covered with many white flowers that face all sides. A hand lens is required to appreciate the flowers.

Eria rigida

This lowland epiphyte of Sumatra, the Malay Peninsula and Kalimantan has canes a metre long bearing numerous inflorescences that carry only a single white flower. The five bracts are a dull crimson.

Opposite: *Eria bractescens*

Top: *Eria braddonii*

Above: *Eria albido-tomentosa*

Eria xanthocheila

This is also a very attractive species distributed throughout Indonesia and the southern half of Peninsular Malaysia. It grows by streams in the lowlands. The stems are 20 cm long with 3 or 4 large leaves at the apex. The inflorescence spreads horizontally outwards from the apex of the stem and bears up to 40 delicate flowers that are creamy yellow, 1.5 cm across and well presented. The plant is easy to grow and is generous with its flowers. It is a very desirable botanical to grow and perhaps with which to hybridise.

Left: *Eria cepiflora*

Above: *Eria rigida*

Opposite, Top: *Eria oblitterata*

Opposite, Bottom: *Eria xanthocheila*

The *Eulophia* Tribe

The *Eulophia* tribe has over 300 predominantly terrestrial species distributed throughout the tropics with a concentration in tropical Africa. The plants have prominent pseudobulbs with a few apical leaves.

Genus: *Geodorum*

This genus of 10 species of attractive terrestrial orchids belongs to the *Eulophia* tribe and is distributed from India to Australia, with only two species represented in northern Peninsular Malaysia. The round pseudobulbs are subterranean, and from these arise a few stalked, folded leaves. The name is derived from the Latin *geo*, meaning 'earth', in reference to its habit.

The scape is erect with the inflorescence nodding just before the flowering rachis, and the half-opened flowers hang downwards, clumped together. This nodding inflorescence is a typical characteristic of the genus. The flowers are delicately coloured but do not last long. According to Steiner and Davis, the tubers of *Geodorum nutans* are used as a disinfectant in the Philippines, and the mucilage extracted from the tubers is used to produce a special gum for gluing guitars in the northern provinces of Luzon.

Geodorum citrinum

Geodorum citrinum is found in Myanmar, Thailand and Peninsular Malaysia. The tepals are pale yellow, almost white, and the flowers are 4 cm across. The foliage is very attractive. The plant illustrated was collected from Pulau Jerejak, a tiny island close to Penang. No hybrids are recorded.

The *Liparis* Tribe

The *Liparis* tribe consists of four genera which are not of particularly outstanding horticultural value, but two genera, *Malaxis* and *Liparis*, appeal to growers who specialise in botanicals. They are mostly terrestrials, but some are epiphytes, and the leaves are often plicate. They all have a terminal inflorescence. The flowers do not have a column foot, nor a spur to the lip; and the anther falls off when disturbed.

Genus: *Liparis*

This is a large genus with a worldwide distribution, being excluded only from the Arctic and the Antarctic. Its 200-member species are made up of terrestrials, lithophytes and epiphytes. They are sympodial orchids with spindle-shaped pseudobulbs. The inflorescence is terminal and erect, bearing numerous small flowers.

The terrestrial species resemble *Malaxis*, but the lip is at the bottom of the flower instead of on top, as in *Malaxis*. The epiphytic species have one or two narrow, tough leaves jointed at the base. Most of the epiphytic species have long scapes with greenish yellow or purple flowers facing all directions like the hyacinth. The flowering is seasonal.

Liparis viridiflora

A beautiful, free-flowering species which can be grown into handsome specimen plants capable of producing a dozen sprays simultaneously, *Liparis viridiflora* enjoys a wide distribution from India and Sri Lanka to the Philippines. It is found on Penang Hill as well as in the lowlands and mountains of Perak and Pahang in Peninsular Malaysia.

The pseudobulbs are flattened, 3–9 cm long with two long leaves, 25 cm by 3.5 cm. The inflorescence is 15 cm tall, crowded with pale greenish flowers. The curious flowers have a dorsal lip which is yellow.

Genus: *Malaxis*

This is a large genus of 300 species with a worldwide distribution, although they are more common in the tropics. They are terrestrial plants (or very rarely lithophytes or epiphytes) with small flowers which are dwarfed by the large plicate leaves. They are unlikely contenders for the attention of the orchid grower, and no hybridisation has been carried out with them.

Opposite: *Geodorum citrinum*

Below: *Malaxis latifolia*

Right: *Liparis viridifolia*

Malaxis latifolia

This orchid is found throughout the lowland forests of Peninsular Malaysia and is widely distributed from India and China to Australia. It grows in heavy shade and flowers frequently. The flowers face all directions and are tightly clustered around the distal half of the erect inflorescence, which measures 15 cm. They are green, deeply flushed with purple lower down the stem, and have a dorsal lip. Frequently, the entire inflorescence is pollinated, producing clusters of seed pods.

The *Nephelaphyllum* Tribe

The name of this tribe means 'cloud-leaf', referring to the clouded variegation in the coloration of the leaves. It is a small tribe of terrestrial orchids with four Malaysian genera, all of which are rather uncommon and very rarely cultivated.

Genus: *Tainia*

Tainia belongs to the rather confused *Nephelaphyllum* tribe and has six Malaysian species, none of which is common. The six other species are distributed from Taiwan to Papua New Guinea. They are terrestrial orchids with pseudobulbs bearing pleated or fleshy leaves and tall, erect inflorescences of well-spaced, handsome flowers of exotic coloration.

No hybrids have been registered with *Tainia*, and its genetic affinities are unknown.

Tainia penangiana

This species is found only on Penang Hill and the Taiping hills in Peninsular Malaysia, but it grows and flowers well in lowland cultivation in Singapore. It is an attractive species with large flowers. The sepals and petals are pale yellow with purple veins, and the lip is white, splashed with yellow on the midlobes. The flowering season is October to February.

Opposite: **Three clones of** *Malaxis latifolia*

Right: *Tainia penangiana.* It is a pity that these beautiful orchids are rarely seen.

The *Odontoglossum* Tribe

This is an enormous tribe that is widely distributed in Central and South America. It is well represented by many beautiful genera (*Odontoglossum*, *Miltonia*, *Oncidium*, *Brassia* and *Rodriguezia*). Only the *Oncidium* and a few *Brassia* grow well in the tropical lowlands. Although none of the species are native to Asia, members of the genera *Oncidium* and *Brassia* have been cultivated in Singapore since the beginning of an interest in orchids in this area. *Onc.* Golden Shower was bred in Singapore before World War II, but Hawaiian hybridisers managed to register the same hybrid as *Onc. goldiana* before Singapore put in its application.

Genus: *Oncidium*

The *Oncidium*, which are better known as the Dancing Ladies, constitute one of the most important genera among amateur growers. The commoner *Oncidium* flowers are easily recognisable by their large, widespread lip, which is usually frilled along the edges and surmounted by a complex callus, the hallmark of the *Oncidium*. Usually, the petals are small and outstretched, like the arms of a dancing girl, and the dorsal sepal represents her head while the lateral sepals are hidden behind the skirt (lip). The flowers are borne in a light airy manner on long, slim racemes, sometimes singly but more commonly in large numbers on large, branching inflorescences. They come in all shades of the rainbow (except blue) with bright yellows, golds and reds predominating the variety in colour. This is a diverse genus with some 70 members which have a wide range of chromosome numbers. Many primary, interspecific hybrids are quite sterile. Nevertheless, *Oncidium* have been successfully bred to other members of the *Odontoglossum* tribe.

Oncidium are sympodial orchids. A large number of species have large, flattened pseudobulbs bearing 1–3 flattened leaves on top. Some members, such as *Onc. cebolleta*, have terete leaves, either erect or twisted and curled, which have earned them the name rat-tailed *Oncidium*. Others have large, thick, rigid leaves and are known as the mule-eared *Oncidium*. A rather separate group are the equitant *Oncidium*, which have no pseudobulbs and whose leaves radiate from the rhizome in the shape of a fan. They are miniature and extremely colourful.

Oncidium are native to Central America, with a distribution that extends from south Florida through Mexico and the Caribbean to the Andes and Argentina. Its members have adapted to a wide range of habitats, from mountainous rain forests to lowland swamps, with many species occurring in open areas as epiphytes on trees or rocks and even as terrestrials in grasslands. *Oncidium* prefer to be

Opposite: *Onc.* Golden Sunset, an equitant *Oncidium*

Below, Left: *Onc.* Gower Ramsey

Below, Right: *Onc.* (Gower Ramsey x *sphacelatum*)

These are the new yellow hybrids with larger flowers that have replaced *Onc. goldiana* var. Golden Shower in the cut-flower market.

Above, Left: *Onc. cebolleta*

Above, Right: *Onc. papilio*

grown on the dry side, and their roots must not be allowed to become waterlogged.

W.W. Goodale Moir of Honolulu dominated the breeding programme of the *Oncidium* in a way that no one else has been able to do for any other orchid subtribe or genus. His work in the breeding of colourful, equitant *Oncidium* is now being continued by other growers in his native Hawaii, and their beautiful hybrids attract much worldwide interest. The principal source of yellow in cut orchid flowers from Southeast Asia is Singapore's own *Onc.* Golden Shower, which was bred in 1936 and is cultivated by the acre in Singapore, Peninsular Malaysia, Thailand and Indonesia. In time, it may be replaced by the newer, larger-flowered *Onc.* Gower Ramsey. *Onc. haematochilum*, *Onc.* Dr Sharagen and *Onc.* Guinea Gold are old-time favourites with larger flowers, but the trend today appears to favour Goodale Moir's compact, equitant *Oncidium*, which flower in tiny 3–5 cm pots. These *Oncidium* are ideal for providing interesting detail in landscaping.

Oncidium ampliatum

This magnificent, warm-growing *Oncidium* produces large, upright, arching and branching inflorescences up to a metre long, bearing hundreds of small, dancing flowers of clear yellow with reddish spots at the bases, of their tiny sepals and petals. The flowers are up to 3 cm across, and the kidney-shaped lip is broad, round and flat. The pseudobulbs are large, 12 cm tall and 10 cm wide, rounded and tightly clustered, with 1–3 leathery, flattened leaves up to 35 cm long.

Its natural distribution ranges from Guatemala, Peru and Venezuela to Trinidad. It grows well in the tropical lowlands, flowering from March to June.

Oncidium cebolleta
(syn. *Onc. longifolium*)

This is a variable species distributed from the West Indies to Mexico and Paraguay. The plants have a stout rhizome with tiny pseudobulbs and terete leaves, 15–60 cm long. The erect inflorescences are borne in spring and reach up to 75 cm in length, and they may be branched. It bears few or many flowers, up to 4 cm across, with a large yellow lip.

Oncidium flexosum

This species is native to Brazil, Paraguay and Uruguay. It flowers mostly in autumn and winter and grows best at 700 m altitude on the equator. However, it will also grow and flower well in the lowlands in Singapore and Malaysia, and the flowers last a long time. The inflorescence is up to a metre long, with numerous side branches which are densely packed with small yellow flowers 2.5 cm in height. There are brown blotches at the base of the sepals and petals.

Oncidium lanceanum

This distinctive, mule-eared *Oncidium* bears flowers which have relatively large-sized petals and sepals and a purple lip. The petals and sepals are yellow marked with blotches of chocolate brown, reminding one of the colour of batik. Each flower is 5–6 cm tall and 4–5 cm across, about 20 flowers being borne in a tight cluster on an erect branching inflorescence which can be up to 50 cm tall.

The pseudobulbs are greatly reduced, and the thick elliptical, leathery, pale green and mottled leaves appear to rise directly from the stout rhizome. The leaves are 3 cm by 10 cm, borne singly, and new vegetative growth appears one month after flower spike initiation.

The species is distributed from Colombia and Venezuela through the Guianas to Trinidad. It grows and flowers well in Singapore, where it is much admired. The Singapore Botanic Gardens made their first *Oncidium* hybrid by crossing this species with *Onc. luridum* to produce the natural hybrid *Onc. haematochilum*, which is extremely hardy and floriferous. *Onc.* Josephine, a back-cross of *Onc. haematochilum* to *Onc. lanceanum*, is a variable hybrid with some clones closely resembling *Onc. lanceanum*.

Oncidium papilio

According to orchid legend, it was *Onc. papilio* that started the orchid craze in England in the 19th century. The butterfly-shaped flowers are borne singly at the end of a slender, 40 cm long rachis which continues to flower for years, with single blooms appearing in succession to replace the ones that have faded. Each flower lasts from several weeks to two months on the plant. The flowers are 12–15 cm tall, with the dorsal sepal and petals erect, stiff, thin and slightly curved terminally, rather like the antennae of a large insect. The lateral sepals are crescent-shaped, curving downwards to surround a large, round lip. The lateral sepals and the lip are a bright yellow, banded with reddish brown, with undulating edges. A well grown plant bears several inflorescences, the delicate flowers hovering above the plant like a flock of butterflies, swaying with the breeze.

The pseudobulbs are nearly round, flattened and about 5 cm in diameter with a single, mottled, dark green leaf measuring 20 cm by 6 cm. It is native to Trinidad, Venezuela, Colombia, Ecuador and Peru and thrives in all wet, tropical lowland areas up to a height of 700 m. *Onc. kramerianum* is a closely related species found at an altitude of 300–1,000 m on the Andes. The two species are brought together in *Onc.* Kalihi, which also flowers well in Singapore. In other hybrids, the unique shape of *Onc. papilio* is lost, and the hybrids are thus considerably less attractive.

Oncidium sphacelatum

Onc. sphacelatum is a robust, pseudobulbous *Oncidium* well suited to tropical lowland cultivation, and it was the perfect choice as a parent for producing free-flowering *Oncidium* for the tropical lowlands. *Onc.* Golden Shower (syn. *Onc. goldiana*), produced by the Singapore Botanic Gardens in 1936 (x *Onc. flexosum*), is one of the most free-flowering orchids in Singapore and is an outstanding contribution to the cut-flower industry.

The pseudobulbs of *Onc. sphacelatum* are 15 cm long, flat and a pale yellow-green, each bearing two or three lingulate leaves about 30 cm long. The inflorescence is arching, up to 50 cm long, with numerous short side branches carrying many closely arranged yellow flowers in one plate. The sepals and petals are yellow, barred with brown and curved backwards; the large yellow lip is oblong.

The species is widely distributed throughout Mexico, Guatemala, the Honduras and El Salvador. It grows well in Singapore and flowers freely when given full morning sunlight. The peak flowering season is November to June, and it is interesting to note that the crossing with *Onc. flexosum* (which flowers in autumn and winter) flowers throughout the year.

Oncidium triquetrum

This dainty dwarf species is an equitant *Oncidium* from Jamaica which flowers mostly in summer but can be almost continuously blooming when mature. A short infloresence of 10–15 cm carries 5–15 flat flowers 2 cm across and 3 cm high. The lip is enormous compared with the rest of the floral segments; it is splashed with red or maroon at the centre on a white background. The horizontal, wing-shaped petals are also white, lightly splashed with red.

Onc. triquetrum is used extensively in hybridising, because its fine form and colour are transmitted to its numerous interspecific hybrids.

Left to Right: *Onc. lanceanum*; *Onc. laceanum/haematochilum* hybrids are plants for landscaping; *Onc. triquetrum*; *Onc. sphacelatum*

Genus: *Brassia*

This Central American genus is closely related to *Oncidium* and is distinguished from the latter principally by its elongated sepals and petals. Although the name Spider Orchid has been applied to this genus, it is best reserved for the Asian *Arachnis*. Many of the larger species are striking and are popular with hobbyists. The plants have large, flattened pseudobulbs and long leaves. The flowers are well carried on an erect, arching inflorescence, and may exceed 30 cm in length in some species. Yellow, green and brown are the common colours.

Brassia Rex, shown here, is the popular representative of this genus in Southeast Asian collections. It is a primary hybrid between *B. verrucosa* and *B. gircoudiana*, and was bred and registered in 1964 by the famous Hawaiian expert on the *Oncidinae*, the late W.W. Goodale Moir.

Colmanara is a complex intergeneric hybrid which has recently found its way into Southeast Asia. It adapts well to the tropical lowland climate and flowers several times a year.

Above: *Brassia* Rex. This robust plant is easy to grow and flower in Singapore and can be depended upon to produce a magnificent display.

Opposite: *Colmanara* Wild Cat (red variety)

The *Phaius* Tribe

The members of the *Phaius* tribe are terrestrial, sympodial orchids with large, decorative plicate leaves, jointed at the base, and a tall, erect inflorescence rising from the leaf axil on the side of the pseudobulb. There are six Malaysian genera, one genus in Sri Lanka and one in tropical America. All have large flowers and can be cultivated as pot plants. *Calanthe* and *Spathoglottis* are popular.

Genus: *Phaius*

This is a genus of 30–40 large terrestrial orchids found in forests and savannah from Africa to Japan and Australia. They have large showy flowers carried on tall, erect inflorescences that open successively. Several intergeneric hybrids have been made with *Phaius*, but they are unknown in Singapore, where only *Phaius tankervilliae* is cultivated. The genus is not explored because it has failed to gain popularity in Southeast Asia.

Genus: *Calanthe*

'Beautiful flowers' is an appropriate name for this lovely genus which inspired John Dominy to produce the very first orchid hybrid in 1856. They are terrestrial plants with 150 species widely distributed from central West Africa across tropical Asia and Australia to Tahiti with a single representative in Central America. The genus is divided into two groups: evergreen with small, corn-like pseudobulbs; and deciduous with large, angular pseudobulbs.

    Calanthe has large, plicate leaves and tall, erect, sometimes arching inflorescences, with few to many delicate but showy flowers. The commonest colour is white. Some are pink, yellow or orange. The petals and sepals are narrow and equal in size, while the lip is

several-lobed and sometimes marked with protuberances or callosities. The characteristic feature of *Calanthe* is the union of the column with the lip.

Calanthe calcarata

This is a cool-growing pink *Calanthe* which has better form than most members of the genus. The leaves are similar to those of *Calanthe veratrifolia*.

Calanthe rubens

This beautiful species with pink flowers is native to Pulau Langkawi and Peninsular Thailand, where it grows on limestone. The flower scape is 50 cm long and covered with fine hair. It carries 6–8 flowers spaced widely apart. The flowering rachis continues to extend

and flower for some weeks. The flowers are 3 cm across and 3.5 cm tall with narrow tepals.

Calanthe veratrifolia

Also known as *Calanthe triplicata*, this is a very showy species with striking white flowers. It is widely distributed from India and China through Southeast Asia to Papua New Guinea and Australia. In Malaysia, it is abundant in freshwater swamp forests but it can also be found growing at 1,600 m. The leaves are 50 cm long and 15 cm wide, plicate and deep or silvery green. The inflorescence is erect with 30–40 white flowers on the scape, each 2.5 cm wide and with a spot of orange on the lip.

With deep green, plicated leaves, this is a beautiful pot plant even when it is not in bloom. It will flower in quite dense shade and the flowers last 3–4 weeks.

Calanthe vestita

This is a very attractive species with white flowers similar to *Calanthe rubens* but with better shape. The pseudobulbs are also larger. It is found on limestone in Myanmar, Thailand, Malaysia, Kalimantan and Sulawesi. It will grow and flower well in Singapore but needs a dry resting period. It has been much hybridised in Europe.

Hybrids of *Calanthe*

Calanthe Dominii is the first man-made orchid hybrid. The name was coined in 1856 by John Lindley to honour John Dominy, who pollinated *Calanthe masuca* with pollinia from *Calanthe furcata*, sowed the seed, raised the plants and achieved the first flowering. Eighty *Calanthe* hybrids were registered before 1945, but after WWII the genus has been largely ignored.

Opposite: *Calanthe veratrifolia*

Top: *Calanthe calcerata*

Below: *Calanthe rubens*

Genus: *Spathoglottis*

Spathoglottis is a genus of terrestrial orchids which has been widely cultivated for a long time in Southeast Asia. It comprises approximately 40 species, distributed from northern India and southern China across Southeast Asia to Australia and the Pacific Islands of Samoa and New Caledonia. In the recolonisation of land which has been laid barren by natural catastrophies, *Spathoglottis plicata* is among the early pioneers. In the famous volcanic eruption of Krakatoa in 1883 where all vegetation on the island was destroyed, *Spathoglotis plicata* was among the three orchids found on the island by Boerlage 13 years later. The seeds of *Spathoglottis* were probably spread by wind from the neighbouring island of Java. Seeds of *Spathoglottis* germinate immediately on planting, rapidly becoming photosynthetic, and early growth is vigorous.

Spathoglottis can be grown in pots but do better when bedded out. They require good drainage and a rich soil. In Malaysia and Singapore, the common practice is to grow them in a mixture of burnt earth and dried manure. The pseudobulbs should be half buried, and the plants need to be sheltered from intense sunlight for a fortnight after planting. Thereafter, they do best in full sunlight; at the very least, full sunlight in the morning. Watering and manuring should be liberal.

When well grown, many species produce tall erect spikes of bright flowers continuously, with a peak of flowering during the dry season. A few species are deciduous.

Spathoglottis affinis

This beautiful golden, dwarf species is found on Kedah Peak at 100 m to 750 m in open rocky country. It grows well at sea level but requires a resting stage of 3–4 months after it has flowered. The pseudobulbs can be divided and repotted after this resting stage, and when well established, the plant should be put into full sun. Each flowering lasts about four months. Many beautiful golden hybrids have been made with *Spathoglottis affinis* using the other parent to give size and vigour and to eliminate the deciduous habit.

Spathoglottis plicata

Spathoglottis plicata is the most widespread species in the genus, and many horticultural varieties have been described. The typical variety has bright purple flowers, 5 cm across, borne on a terminal cluster on a long, upright 60 cm scape. The lip of the flower is dark purple, and at the base of the midlobe, there are two yellow calli. The flowers open successively, 5–6 flowers at a time, and each spray may have up to 40 flowers. The beautiful pure white variety discovered by Dr. Yeoh Bok Choon, called the 'Penang White', breeds true from seed and has produced a white lineage.

Hybrids of *Spathoglottis*

Spathoglottis Primrose was the first Singapore-bred orchid. It is the quickest to flower from fertilisation. How Yee Peng, his son Wai Ron and, recently, Dr. Tim Yam are producing some fine hybrids.

Opposite: *S.* Premier

Right, Clockwise from Top Left: Two clones of *S. plicata* (left and right); *S.* Primrose (pink); *S. kimballiana*; *S.* (*plicata* x *tomentosa*) x *plicata* (white); a hybrid of *S. affinis* (yellow)

Chapter 5
Monopodial Orchids

Grown by the acre in full sun, the monopodial orchids are all members of the *Vanda-Arachnis* tribe, the single tribe constituting the subfamily *Vandoideae*. They are the most widely cultivated orchids in the world.

The *Vanda-Arachnis* tribe is principally confined to the Old World tropics extending into Madagascar and northern Australia. It comprises some 30–40 genera, including some highly evolved genera that are completely leafless (such as *Taeniophyllum* and *Chiloschista*). The names of its popular member genera read like an honours list of orchids which are distinguished by their showy flowers: *Vanda, Phalaenopsis, Renanthera, Rhynchostylis, Aerides, Ascocentrum, Doritis, Vandopsis* and *Angraecum*. Several genera are interfertile. In some instances, the relationship between two genera appears to be closer than that between two species in the same genus. For instance, strap-leaf *Vanda* is genetically closer to *Ascocentrum* than to terete *Vanda*. A large number of multi-generic hybrids have been raised, their creation adding greatly to the wealth of colour, shape and size of cultivated orchids. Several outstanding species of this tribe have left their personal hallmark on the orchid hybrids — red from *Renanthera*, white from *Phal. amabilis*, blue and roundness from *Vanda sanderiana*, toughness and a long vase life from *Arachnis*.

The characteristic feature of the monopodial orchid is the stem which has an unlimited apical growth and which is enclosed within the leaf sheaths, usually rooting at intervals almost to the top, although sometimes the roots are only formed at the base. The leaves may be flat, pencil-shaped (terete) or intermediate. The inflorescences are produced at the leaf axils, piercing the leaf sheaths as they emerge. The flowers are extremely varied, generally of heavy substance, and contain two or four waxy pollinia. The vegetative form is very similar, and there are no pseudobulbs. However, there is a great variation in the size of the full-grown plant, from the city *Taeniophyllum*, which may be held in the palm of one's hand, to the tall-flowering *Arachnopsis* and the *Renanopsis* Lena Rowold, which bloom at a height of 4 m.

In the tropics, the monopodial orchids are among the easiest plants to cultivate as they will adjust to diverse growing conditions and will accept a wide range of containers and potting mixes. While only certain hybrid types will be perennially in bloom, all monopodial orchids grow continuously throughout the year; and therefore their basic essentially requirements are high humidity at all times, adequate light and heavy fertilising. If compost is heaped around the base of a monopodial orchid that is grown in full sun, the roots which are covered by the compost will branch extensively and their protective velamentous covering will thin out, creating a massive absorptive surface for the plant. This is the key to growing a robust monopodial orchid.

Genus: *Acampe*

Vegetatively, this resembles a robust *Vanda*, and under cultivation, it may be grown like a *Vanda*. It is dispersed throughout tropical Asia and Africa but has not received much attention from botanists or orchidists.

Opposite: *Vanda* Chai Goodfellow x *Ascda.* Nopporn Gold

Above: *Acampe longifolia*

Acampe longifolia

This is found from Sikkim in India to northern Peninsular Malaysia (Penang and Pulau Langkawi) and also in Papua New Guinea. The inflorescence has 1–2 branches and carries up to 25 blooms, but the flowers are bunched together and do not open widely. The petals and sepals are a bright yellow barred with crimson. For its own sake, *Acampe longifolia* is hardly worth cultivating. Two hybrids have been made, one with *Vanda* and the other with *Arachnis*, and both are colourful and very floriferous, although they are seasonal bloomers. More intergeneric crossings should be made with this promising species, perhaps with *Renanthera*.

Genus: *Aerides*

This is a popular genus commonly called the Foxtail Orchid because its beautiful, cylindrical sprays of flowers so closely resemble a fox's tail. There are some 60 species of *Aerides*, of which about 10 are in cultivation. Eight of the cultivated species hail from Thailand. In habit, the *Aerides* are strong monopodial plants which are air-loving, hence their name. The waxy, fragrant flowers are produced in long, dense clusters and open almost all at once. There is a wide range of colours to choose from, such as green, mauve, magenta, yellow and white. The flowers have spreading petals and sepals, with the lateral sepals broader than the dorsal. The curious beak-like spur formed by the column foot and the lip is the most characteristic part of the flower.

Foxtail Orchids are easily grown into superb specimen plants, and in the tropics they can be grown tied to garden trees. Even when they are completely ignored, they will grow into large clumps. Nowadays, they are commonly grown in empty wooden baskets. If they are tied to a tree, the stems will hang downwards with the growing part recurving towards the light. The stem branches freely whenever it bends, and gives off aerial roots which reach back to the supporting tree. In baskets, the plants naturally grow upwards and they will only produce offshoots at the base of the plant. *Aerides* require a sunny location and free air movement. They like frequent application of fertiliser. When they are well grown, they are most spectacular, producing 10, 20 or 50 sprays of flowers all opening at the same time. The peak flowering period for *Aerides* in the north of Thailand is April to May.

Several species of *Aerides*, such as *Aer. odorata*, *Aer. falcata* and *Aer. houlettiana* are fragrant. This fragrance was much admired by the nobility in feudal Japan.

Extensive intergeneric hybridising has been carried out with *Aerides*. Several hybrids are spectacular; all are free-flowering and floriferous. The characteristic of simultaneous opening of the flowers is also transmitted to the hybrids.

Aerides houlettiana

An attractive species from northern Thailand and Indochina, some authorities consider it to be a variety of *Aer. falcata*. The inflorescence is short, but it carries 12–20 well arranged, pale yellow flowers with a wedge of lavender at the lip. The flowers are faintly fragrant.

Aerides lawrenceae

Perhaps the most admired species in the genus, this occurs in the Philippines in the southern island of Mindanao. It has larger, rounder flowers than *Aer. odorata* and is commonly used as a hybrid parent. The standard form has white sepals and petals that are marked with crimson at the tip. The flowers are pollinated by large bees. The column is hidden by the two sidelobes and the midlobe of the lip; the bees have to work their way between the hinged lip and the column in order to reach the nectar in the spur. During this struggle, they dislodge the pollinia, which get stuck to their heads and are borne to the next flower.

An albino variety known as *Aer. lawrenceae* var. *fortichii* has been collected in the Bukidnon province of northern Mindanao, but it is rare. The sepals and petals are white and the spur is tinged with green.

Aerides multiflora

This is a very beautiful dwarf species found in the Himalayas southwards to northern

Opposite: *Aerides* Amy Ede (*lawrenceae* x *jarckiana*)

Above: *Aerides multiflora*

Right: *Aerides lawrenceae* (close-up)

Peninsular Malaysia. It has been found in diverse locations. The inflorescence is sometimes branched at the base and carries 50 more closely spaced flowers during April and May. The flowers are 3 cm in diameter and are white suffused with purple towards the tip. The lip is shaped like a spade and is purple with a deeper shade towards its centre.

Aerides odorata

This is a highly variable species which is widely distributed throughout Southeast Asia. It flowers around May, producing numerous sprays, each with 30 fragrant flowers, all opening simultaneously. The common type has white sepals and petals which are blotched with violet at the tip. The prominent spur is spotted with purple. An *alba* variety is very much sought after. A well-grown plant of *Aerides odorata* is a beautiful sight. The flowers last about a fortnight.

According to Kamemoto and Sagarik, the plants from northern Thailand are tetraploid and carry 50 flowers on each spray. They would be very desirable for hybridising.

Aerides rosea

Native to Myanmar, northern Thailand and Laos, this species requires a distinct dry season to produce long sprays of overlapping, rosy pink flowers featuring a round lip of a darker pink. There are usually 20–25 flowers on an inflorescence, but selected clones have more flowers.

Hybrids of *Aerides*

A large red lip with accent on the spur, close arrangement, flat flowers and a rather delicate suffusion of colour on the sepals and petals are the four characteristics that one may expect *Aerides* to impart to its hybrids. A large lip and close arrangement with numerous flowers characterise all hybrids of *Aerides*. Those hybrids which have inherited a brilliant colouring from the other parent have shown the *Aerides* lip to greatest advantage, as in *Aeridopsis* Teoh Phaik Khuan (*Paraphalaenopsis denevei* x *Aerides lawrenceae*) (see *Paraphalaenopsis*).

Since *Aerides lawrenceae* is the finest species, it is the one which is more frequently used in hybridising. *Aerides* has been bred successfully to strap-leaf and terete *Vanda*, *Arachnis*, *Ascocentrum*, *Ascoglossum*, *Neofinetia*, *Phalaenopsis*, *Paraphalaenopsis*, *Renanthera* and *Rhynchostylis*. A number of multi-generic hybrids have also been raised (*Burkillara*, *Christieara*, *Carterara*, *Lewisara*, *Lymanara*, *Nobleara*, *Perreiraara*). Most of these are free-flowering. The *Aeridovanda* have beautiful, large flowers of a delicate complexion, though generally without much variation in their colour, an exception being the beautiful yellow *Aeridovanda* Fuchs Yellow Jacket var. MGR's Gold Coin AM/AOS. *Aeridopsis* Teoh Phaik Khuan is vivid because of the contribution from the *Paraphalaenopsis denevei*. *Aeridarchnis* Bogor 'Apple Blossom', made from the *alba Aerides odorata*, is much admired because it is a white scorpion-type

orchid. Thai growers have used their mastery over the vandaceous genera to impart a range of colour to the hybrids of *Aerides* and simultaneously improve on the size of the flowers. While they too prefer using *Aerides lawrenceae* as a parent, they have not ignored their own local *Aerides coelestis* and *Aerides flabellata*. *Aerides coelestis* yields a more compact and upright inflorescence with plenty of flowers, but typically, the colour is faint. Hybrids of *Aerides flabellata* acquire compactness. *Aerides* are tough, vigorous plants and so are their hybrids.

Opposite: *Aerides lawrenceae* x *Ascda.* Kultana Gold

Right: *Aeridorachnis* Bogor var. 'Apple Blossom'

Below, Left: *Christieara* Luang Prabang

Below, Right: *Aeridovanda* Suruko Iwasaki

Opposite, Clockwise from Left: *Aeridovanda* Veravoon; *Christieara* George Heider (*Aer. lawrenceae* x *Ascda.* Duang Porn); *Aerides lawrenceae* x *V.* Manuvadee

Left: *Vascostylis* Nong Kham x *Rhy. coelestis*

Above: *Aer. lawrenceae* x *Ascda.* Guo Chai Long

Genus: *Arachnis*
The Scorpion Orchids

Sun-loving orchids, the *Arachnis* and its numerous intergeneric hybrids are certainly the easiest orchids to grow in the tropics. Throughout Singapore and Malaysia, most homes which have gardens will own a few pots of these orchids. In the monotonous climate of the equatorial belt, where spectacular flowering of garden plants is the exception rather than the rule, the Scorpion Orchids have earned a place for themselves in the garden because they constantly produce large numbers of gorgeous long-lasting sprays. In commercial nurseries in Peninsular Malaysia and in Singapore, they are grown by the acre in full sun to provide the millions of orchid blooms demanded by the florist trade. They have no equal for hardiness, ease of propagation, floriferousness and flower durability. Today, *Aranda* has replaced the historic *Vanda* Miss Joaquim as the representative orchid from Singapore.

The genus *Arachnis* is popularly known as the Scorpion Orchid because its flowers resemble a scorpion, the curved lateral sepals simulating the claws and the dorsal sepal resembling the tail. At one time, it was included in the genus *Renanthera* to which it has a likeness in its vegetative appearance. The monopodial *Arachnis* has long, climbing stems which branch whenever they are bent. The leaves are oblong and narrow, tapering towards the tip, where they become bilobed. They are tough and leathery and will not wilt when exposed to the full blast of the tropical sun, even if they are not watered for several days. There are 15 different species, but few are in cultivation. Some have a short inflorescence with few flowers, while others have long, erect or pendulous sprays over a metre long, carrying large flowers which are set rather far apart. The petals and sepals are of equal size, long and narrow, clubbed at the tips and with the edges curved backwards. The lateral sepals and petals are curved downwards at the tips. The column is short and stumpy. The tip is relatively small and is attached by a hinge to a thick column foot.

Arachnis is distributed from Myanmar and Indochina through Malaysia and Singapore into Indonesia and Papua New Guinea. There are only three species in Peninsular Malaysia, one being the natural hybrid of the other two. However, these three species are the most widely cultivated and account for 99 percent of the hybrids with this genus.

Nowadays, species plants of *Arachnis* are not commonly grown, because their hybrids surpass them in brilliance, colour, floriferousness and even in cost. The traditional method of planting these orchids in beds in full sunlight is still the most convenient way to grow *Arachnis*. Two rows of posts 60 cm tall or two rows of horizontal wooden beams 30 cm by 60 cm above the ground may be used to anchor the plants, whose stems should be about 10 cm above the ground. Wood shavings, cut grass, particularly *lallang* (*Cylindrica imperata*), or garden compost are piled around the base to shelter the feeding roots, which will grow into the compost when they ramify extensively. When newly planted, they need partial shading from direct sunlight, but after a few weeks the shading can be removed as the established plants will grow best when they receive the maximum amount of daylight. It is important to realise that the supports should not be too high, as they will

Opposite: *Arachnis longisepala* is a newly discovered species from Sabah, Borneo.

Right: *Aranda* Baby Teoh (*Aranda* Christine x *V.* Somboon) typifies the larger second-generation *Aranda*.

not flower until a good portion of the stems have grown free above the support. However, as the plants become lanky, the supports may be extended. *Arachnis* like heavy fertilising and being watered daily, or even twice a day during the dry season, the exception being *Arachnis lowii*, which needs to be kept dry. Propagation is by top-cutting.

Arachnis breviscapa

This beautiful species occurs in Borneo and flowers only once a year in Singapore, at the end of the wet season. The inflorescence is short (5–6 cm) and carries only 2–6 flowers, but the sepals and petals are broad and the colour is clear. The basal colour is cream with large, irregular, brown markings. The plant likes a little shade. *Arach.* Capama, a cross between *Arach. breviscapa* and *Arach.* Maggie Oei, is an exceptionally beautiful, yellow Scorpion Orchid, but like its first parent, it is a seasonal bloomer.

Arachnis flos-aeris

This is the typical Scorpion Orchid that is more akin to its animal namesake than any other species of *Arachnis*. It is widely distributed in Sumatra, Java and Borneo and is found on limestone in Perak and Pahang in Peninsular Malaysia, although uncommon. The plants are large and rambling, with leaves 15–20 cm long and 5 cm across. The inflorescence is up to 150 cm long, branched, bearing large flowers 10 cm tall and 8 cm across, and spaced widely apart. The sepals and petals are a pale yellow-green marked with broad, irregular bars of purplish brown. The flowers of *Arach. flos-aeris* give off a strong musk, enabling the orchid hunter to track down the plants. They are difficult to locate when not in bloom. Previously, it was also known as *Arach. moschifera*.

In addition to the type, there are two varieties, the better known being the Black Scorpion Orchid, *Arach. flos-aeris insignis*. The petals and sepals of the Black Scorpion Orchid are a dark maroon, almost black towards the tip, and the texture is waxy. The lip stands out as a patch of white in the centre of the flower. When the plant is not in bloom, this variety can be recognised by its young leaves, which are flushed with purple. It is a native of Sumatra but is cultivated in Singapore and Peninsular Malaysia.

The other variety, *Arach. flos-aeris gracilis*, has flowers which are about four-fifths the size of the type while its sepals and petals are more curved. Its scent is distinctly unpleasant. It is a native of the mangrove swamps in Selangor and Negeri Sembilan on the Malay Peninsula.

Arach. flos-aeris is a seasonal bloomer. The type form flowers twice a year in May and November, while the varieties *insignis* and *gracilis* flower once a year in June.

Arachnis hookeriana

This is a coastal plant which grows on scrubland 100–500 m from the beach along the eastern coast of Peninsular Malaysia and formerly on the offshore islands of Singapore. It is found in Borneo and in the Rhio Archipelago which is south of Singapore.

The plant is compact with leaves 9 cm long and 2 cm wide. It bears an erect inflorescence, usually unbranched, 60 cm long with 6–8 widely spaced, creamy white flowers, each 6.5 cm tall and 5.5 cm across. The petals and sepals are narrow and only slightly expanded at the tip. A well grown garden plant will carry 3–4 sprays at a time with 10–15 flowers on each spray.

There are three varieties, the commonest being the variety *alba* (the White Scorpion Orchid), which is not a true *alba* form at all.

The sepals and petals are spotted purple in their mid-portion, but the distinctive feature is the purple lip which is marked with six dark purple stripes. The variety *luteola* has no purple colouring on the flower and the petals and sepals are more yellow towards the tip while the lip is a pale yellow. The third variety, *viridipes*, has a bright green scape and spotless petals but the lip is purple and marked with six purple stripes.

Arach. hookeriana alba is free-flowering in Singapore, and up to the mid-1970s, it was still widely cultivated for cut flowers. The variety *luteola* is favoured by hybridisers who recognised that its hybrids are more free-flowering, the colours of the hybrid blooms are sparkling, and the inflorescence has a better arrangement.

Arachnis longisepala

This is a new species from Sabah and was exhibited by the Sabah Orchid Society at the Singapore Orchid Show in 2003. It is a tall, scrambling plant bearing long, pendulous inflorescences and lanky, wiry flowers that are spaced far apart. The basal colour of the flower is greenish yellow, overlaid with blotches of deep purple along the entire length of the dorsal sepal and the proximal half of the other sepals and petals. The lip is white or a pale yellow. The flowers are 10–11 cm tall and 7 cm across, with dorsal sepal of 7 cm length and 5 mm width.

Arachnis lowii

This giant among the Scorpion Orchids has been extensively collected from Sarawak, where it occurs on trees overhanging rivers. It is an enormous plant and presents a spectacular sight in full bloom, throwing long, limp inflorescences which are each up to 3.5 m long. The first 1–3 flowers are dark

maroon. In nature, seed pods are usually produced on the yellow flowers. It is not commonly cultivated in Singapore because it flowers only once a year and also because of its size. John Seal described several specimen plants being grown on rambutan trees by Ong Hup Leong in Kuching. Holttum recommends that it be grown either in large tubs or on a coral rockery in light shade with strong wooden posts for support. It should be heavily fertilised during the rainy season but requires a dry season without feeding in order to flower.

It is now reclassified as *Dimorphis lowii* because of the presence of two distinct types of flower on the same spray.

Opposite, Top: *Arachnis hookeriana* var. *luteola*

Opposite, Bottom: *Arachnis hookeriana alba*

Below: *Arachnis breviscapa* — currently reclassified as *Vandopsis breviscapa*

Arachnis maingayi

A natural hybrid between *Arach. hookeriana* and *Arach. flos-aeris*, this species is found near the coast of southern Peninsular Malaysia, on the islands south of Singapore and in Borneo. It is called the Pink Scorpion Orchid because the base colour of the upper sepal is a solid pink, and the petals and sepals are marked with light salmon pink bars. Quite free-flowering, it was at one time commonly cultivated in Singapore, but it has now been completely replaced by an identical artificial hybrid, *Arach.* Maggie Oei. This artificial hybrid was produced by using

the *luteola* variety of *Arach. hookeriana*. Three clones received HCC/MOS in 1961. It never stops flowering in Singapore while a similar crossing involving *Arach. hookeriana* var. *alba* simply refused to flower in the same place.

Hybrids of *Arachnis*

Singapore's orchid trade and the popularity of the Scorpion Orchids owe their origin to Professor Eric Holttum and John Laycock, who produced the first free-flowering *Aranda* (Deborah) and the first free-flowering hybrid *Arachnis* (Maggie Oei) respectively. As so often happened with hybrids of that era (cf. Van Brero's *Vanda* Josephine van Brero and Deventer's *Vanda* Emma van Deventer in Java), *Arachnis* Maggie Oei var. 'Red Ribbon' proved to be tetraploid, and it has been used (at first, without knowledge of its polyploidy) to produce numerous secondary hybrids, some of which were quite outstanding (such as *Arach.* Capama, *Aranda* Tyersall, *Aranda* Majula and *Aranda* Neo Hoe Kiat). *Arach.* Ishbel (*maingayi* x *hookeriana*), produced by the Singapore Botanic Gardens in 1940, was not spectacular in itself but was free-flowering and fertile. Its hybrids were almost always free-flowering and it was soon crossed with a large number of *Vanda, Renanthera, Paraphalaenopsis, Ascocenda, Vandopsis* and *Arachnis* itself. Among the *Aranda* derived from *Arach.* Ishbel, at least three are amphidiploid (and thus fertile) — *Aranda* Tay Theng Suan (produced by How Yee Peng), *Aranda* Bintang Raffles var. Raffles Park AM/MOS (1962) (bred by Chong Chok Chye) and *Aranda* Lily Chong (bred by the Singapore Botanic Gardens).

Aranda Lucy Laycock (*Arach. hookeriana* x *Vanda tricolor*) was produced by John Laycock in the 1950s, and it too was amphidiploid, perhaps the first *Aranda* found to be fertile. *Aranda* Lucy Laycock is the parent of several outstanding hybrids, among them the red

Aranda Kooi Choo, the fragrant *Aranda* Peng Lee Yeoh and the spectacular *Holttumara* Loke Tuck Yip var. Prachuab AM/OSSEA (1968).

The flowering characteristics of these early crosses were valuable object lessons for the Singapore-Malaysian hybridiser, who learnt that *Arach. hookeriana* var. *luteola* was the preferred parent, over all other varieties of *Arach. hookeriana*. Hybrids of *Arach. flos-aeris* were seasonal. *Arach.* Maggie Oei, so free-flowering itself, produced hybrids that were less free-flowering than those produced by *Arachnis* Ishbel. Additionally, the hybrids obtained from *Arachnis hookeriana* var. *luteola* and those produced with *Arach.* Ishbel were cleaner, and the seedlings from a single pod of simple crosses were fairly uniform.

With this knowledge at hand, a succession of fine *Aranda* were produced after the war, each superceding the earlier one, such as *Aranda* Wendy Scott, *Aranda* Christine, *Aranda* Yvonne Tan, *Aranda* Tay Swee Eng and *Aranda* Singapura. The discovery of a tetraploid, blue *Vanda* Dawn Nishimura, in the late 1960s brought about the next generation of giant-sized, blue-violet *Aranda* (*Aranda* Noorah Alsagoff, *Aranda* How Yee Peng, *Aranda* Ang Hee Seng and *Aranda* Wong Bee Yeok), which were equally fine. *Aranda* Christine '80' has produced two of the loveliest hybrids to serve the cut-flower trade — the intense, imperial purple *Mokara* Mak Chin On and the shapely, large, pink *Aranda* Baby Teoh. The last two hybrids are very floriferous and flower at very short height.

Fifteen years ago, certain colour ranges (yellow, green, red and white) were lacking in the present selection of cultivated *Aranda*. An extremely fine *Aranda* had been produced by crossing *Arach. hookeriana* with *Vanda denisoniana*, and the author expressed the hope that perhaps we shall soon see a green *Aranda*. That has yet to come, but in the meantime,

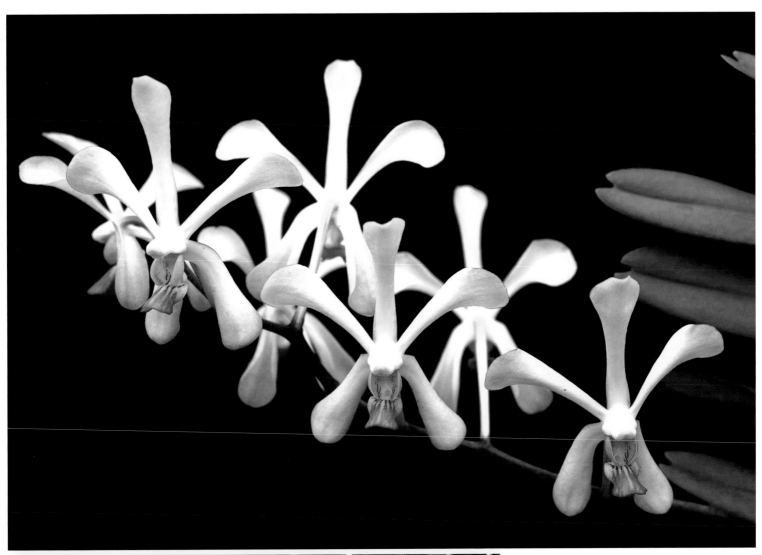

Top: *Aranda* Nancy (*Arach. hookeriana* var. *luteola* x *V. dearei*)

Left: *Arachnis* Maroon Maggie (*Arachnis* Maggie Oei x *Arachnis flos-aeries insignis*) var. Cecilia shown by Dr. E.C. Pink awarded AM/MOS in 1963

Opposite: *Arachnis* (*Dimorphis*) *lowii*

Right: *Arnth.* Anne Black var. 'Mandai' AM/MOS (1962) is still a favourite commercial cut flower. It is easy to cultivate and bloom. Its sprays are manageable, easy to pack and ship.

Below: *Arnth.* Gracia Lewis var. 'Burong Merah' AM/MOS (1964) is by far a more spectacular plant in the garden. The enormous spray is over a metre long and has 3–5 side branches carrying around 100 flowers. Its size makes it difficult to handle as a cut flower and, for this reason, the plant is rarely seen.

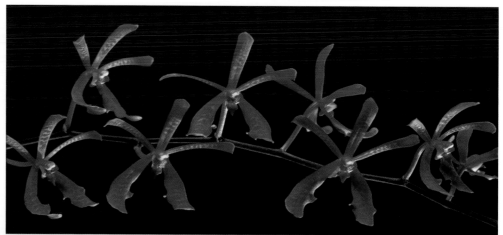

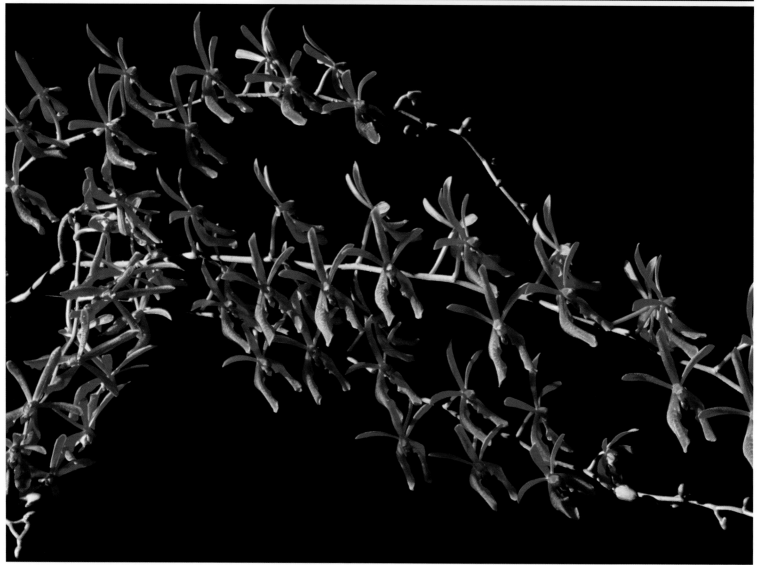

a very wide range of extremely floriferous *Mokara* with large, broad-petalled flowers in sparkling yellow have been produced by breeding *Arachnis* to several *Ascocenda* carrying the parentage of *V. denisoniana* and *V. dearei* for colour and *V. sanderiana* for shape and size (*Mkra.* Singa Gold, *Mkra.* Bangkok Gold, *Mkra.* Lion's Gold, *Mkra.* Zhu Rongji, *Mkra.* World Trade Organization and *Mkra.* Zaleha Alsagoff). Yellow is decidedly the 'in' colour of Scorpion Orchids in the early 21st century.

The classic *Arnth.* Beatrice Ng var. 'Conference Gold' AM/MOS (1963) and the yellow *Arnth.* Gracia Lewis are both very beautiful but scarce. *Arach.* Capama (*Arach.* Maggie Oei x *Arach. breviscapa*) was a remarkable Scorpion Orchid with broad petals and sepals, heavy substance, and clear colours; but despite winning four Awards of Merit from the Malayan Orchid Society, it became extinct because local growers simply did not have space for a huge plant which bloomed for only one month in a year. Another of the author's favourites, *Aranda* Punctata var. City of Singapore AM/MOS (1962), met a similar fate.

Free-flowering Scorpion Orchids have been produced by crossing *Arachnis* with *Renanthera* (*Arnth.* James Storie, *Arnth.* Anne Black., *Arnth.* Bloodshot, *Arnth.* Beatrice Ng, *Arnth.* Gracia Lewis), and some are widely cultivated for their cut flowers. For the hobbyist, some of the polygeneric red hybrids

Four *Aranda* from the 20th century

Clockwise from Top Left: *Aranda* Deborah, the very first *Aranda*; *Aranda* Christine, a name in the 1970s; *Aranda* Majula, very colourful but short spray; *Aranda* Noorah Alsagoff, a second-generation *Aranda* from the tetraploid blue *Vanda* Dawn Nishimura, registered in 1972 and still a major cut flower in the early 21st century

are outstanding, such as *Holttumara* Maggie Mason 'Crimson Spray' AM/MOS (1963), *Holtt.* Park Nadeson, *Sappanara* Ahmad Zahab and *Bokchoonara* Khaw Bian Huat.

The most widely cultivated White Scorpion Orchid is *Aeridarchnis* Bogor, a Hawaiian hybrid. Its small, delicate, strongly scented flowers are spaced rather far apart. The *Mokara* (*Arachnis* x *Ascocenda*) represent an attempt to miniaturise the Scorpion Orchid plant (not merely inflorescence and flower size) and make it even more free-flowering; the creation of new colours is the third objective. All *Mokara* awarded by the Orchid Society of Southeast Asia (such as *Mkra.* Mak Chin On, *Mkra.* Khaw Phaik Suan and *Mkra.* Princess Mikasa) are crossings between large-flowered, second, third or later generation *Ascocenda* and *Aranda*.

On these two pages, from Left to Right: *Mkra*. Singa Gold var. Pamelia (*Arach. hookeriana* x *Ascda. Bangkhuntian Gold*) has a solid golden colour; *Mkra. World Trade Organization* (*Mkra.* Khaw Phaik Suan x *Ascda.* Fortune East) is beautifully speckled in the tradition of the older Spider Orchids; *Mkra.* Bangkok Gold (*Arachnis hookeriana* var. *luteola* x *Ascda.* Mem. Kenny) with smaller, but radiant flowers; *Mkra.* Zhu Rongji (*Arach.* Maggie Oei x *Ascda.* Guo Chia Long), a famous VIP orchid named by the Singapore Botanic Gardens for the Premier of the People's Republic of China to commemorate his visit to Singapore

Genus: *Ascocentrum*

Ascocentrum is a small genus closely resembling *Vanda* and breeding freely (though not naturally) with the latter genus to produce the popular 'miniature *Vanda*', which are known as *Ascocenda*. The plants themselves have the appearance of small strap-leaf *Vanda*. The inflorescence is erect, bearing many small flowers which face all directions. Each flower is flat, brightly and evenly coloured, and has a long spur which points vertically downwards. The individual species present a spectacular sight when they are in bloom; even the small plants bear multiple inflorescences packed with numerous flowers.

Two species occur in the northern Peninsular Malaysia, with one of them, *Asctm. miniatum*, extending into Thailand and Myanmar, where two more species are found. They are epiphytes on trees in deciduous forests. In cultivation today, they are commonly grown in teak baskets which are suspended in lath houses with 25–50 percent shade, whereas in the past, they were kept in clay pots.

Ascocentrum ampullaceum

This species is distributed across Laos, northern Thailand and Myanmar into southern China and Bhutan. It requires a long, cool, dry period to initiate flowering. The 20 (or more) mauve flowers, 1.2–1.5 cm across, borne on a short raceme, are produced in April on a compact plant.

Left: *Ascocentrum miniatum*

Opposite: *Ascda.* Yip Sum Wah, bred by Roy T. Fukumura of Hawaii. The cross received more than 80 awards, including FCC/AOS and FCC/HOS. It is in the parentage of hundreds of modern hybrids.

Ascocentrum curvifolium

This is the biggest member of the genus. Its best clones have bright red, round and overlapping flowers, and it is understandable that it has played a prominent role in the breeding with *Ascocentrum*. It is a montane species from the northern and western parts of Thailand and adjacent regions of Myanmar. It does not flower well in Bangkok and not at all in Singapore. Its hybrids, however, adapt to a wide temperature range, being equally at home in Singapore or in California, and they never seem to cease flowering, in some cases producing flower spikes at every axil once the plant is mature.

Asctm. curvifolium has long, recurved, light green leaves which are spotted with purple along the edges. The plants commence flowering when they are 15 cm tall and may produce 6–7 sprays when they reach a height of 60 cm. Each inflorescence is 15–20 cm tall and erect, bearing evenly coloured, deep orange, red or reddish-brown, round flowers, 2–2.5 cm in diameter. The petals and sepals are round and overlapping, producing a very full form which is highly desirable in orchids. The flowering season is March and April. Individual flowers last a fortnight.

Ascocentrum hendersonianum

This is a delicate, petite, dwarf *Ascocentrum* from Borneo. It has few leaves, each up to 12 cm long. The inflorescence is erect, arising from the base of the stem, 15 cm tall with up to 30 flowers. The flowers are flat, 3 cm across and of a brilliant magenta rose. The lateral sepals are long and oval. It is easy to cultivate and flowers freely in Singapore.

Ascocentrum miniatum

This species has a wide distribution from the Himalayas to Java and is extremely common in Thailand, where it is found between 300 m and 800 m altitude. In Peninsular Malaysia, it is only found in Perlis and Pulau Langkawi. It occurs in deciduous forests at exposed locations.

Although the flowers are much smaller than those of *Asctm. curvifolium* (they are about 1.5 cm in diameter), the plant itself is so much easier to grow, adapting well to diverse situations. The flowers are bright orange with a small black dot in the centre (the anther cap is black) and are tightly arranged on a rather short inflorescence. Orange is dominant in the hybrids, even overpowering the red of *Asctm. curvifolium* when the two species are interbred. It is a delight to see small plants, barely 6–8 cm tall, already producing their maiden blooms of orange flowers. Large plants will produce many sprays which are closely clustered around the stem.

Ascocentrum auranticum is a golden orange species with small flowers, 9 mm across, with up to 40 flowers on an inflorescence. It is rather similar to *Ascocentrum miniatum* and may be the Philippine or Sulawesi strain of the latter species. It flowers during the rainy season in January and February.

Hybrids of *Ascocentrum*

'Miniature *Vanda*', as hybrids of *Ascocentrum* are popularly called, took the orchid world by storm in the late 1960s. These first-generation *Ascocenda* had all the desirable qualities, such as hardiness, a dwarf habit, the ability to flower continuously, upright sprays bearing numerous dainty, long-lasting, well arranged, round, flat flowers in brilliant hues of red, orange, yellow, lilac and blue, and some with tessellations. However, as back-crosses were continued with

Clockwise from Top: *Ren. monachica* x *Ascda.* Fuchs Gold; *Ascda.* Suksaran Sunlight; *Ascocentrum ampullaceum*

Left: *Mkra.* Zaleha Alsagoff 'Nong' FCC OSSEA (1997). This orchid received the only First Class Certificate ever awarded by the Orchid Society of Southeast Asia (and its predecessor, the Malayan Orchid Society). (Photo by courtesy of Syed Yusof Alsagoff)

Below: *Ascovandoritis* Lion's Doll. This is a complex trigeneric hybrid that is constituted by *Ascocenda*, *Vanda* and *Doritis*. The *Ascocenda* has given it form, colour, miniaturisation and floriferousness. The *Doritis* produced the long, upright inflorescence. The *Vanda* has imparted a better arrangement and substance but not a substantial size.

Vanda, the plants and flowers became larger, and in many cases, these latter generations of *Ascocenda* lost the charm of the original dainty *Ascocenda*. The great names among the early *Ascocenda* are Meda Arnold, Ophelia, Eileen Beauty, Yip Sum Wah, Medasand, Memoria Choo Laikeun, Queen Florist and Peggy Foo. One sees such names in the pedigree of many hybrids.

A great asset of the *Ascocentrum* is its ability to add brilliance and intensity to the colour of its hybrid. These are essential yardsticks when assessing the worth of the hybrid. Miniaturisation of *Renanthera* was achieved through the creation of the solid red *Kagawaara* (*Ascocentrum* x *Renanthera* x *Vanda*), and compact Scorpion Orchids (the plants flowering at short height) have arrived with the appearance of *Mokara* (*Arachnis* x *Ascocentrum* x *Vanda*). *Doritis* acquired an orange coloration when bred with *Asctm. miniatum* (as in *Doricentrum* Pulcherrimin and *Beardara* Charles Beard) which has also imparted an apricot coloration to *Phalaenopsis* (in *Asconopsis* Irene Dobkin).

The breeding of *Ascocenda* to *Aerides* produced *Christieara*, a very pleasant combination which expresses many of the desirable features of the parents. The awarded hybrids were bred from an outstanding *Ascocenda* (such as Peggy Foo or Yip Sum Wah) with *Aerides lawrenceae*. *Christieara* Michael Tibbs var. Redland (*Aer. lawrenceae* x *Ascda.* Peggy Foo) was awarded a First Class Certificate by the American Orchid Society.

Vascostylis is the combination of *Ascocentrum* with *Rhynchostylis* and *Vanda*. Here the predominant species is *Rhynchostylis coelestis*, which maintains the erect, multi-flowered, compact inflorescence and manageable size, and simultaneously imparts its delicate blue coloration. Several *Vascostylis* have garnered awards from the American Orchid Society.

On these two pages, from Left to Right: *Ascda.* Lion's Sunbeam; *Vascostylis* Prapin x *V. luzonica*; *Ascocentrum miniatum* x *Ascda.* Bangkhuntian Gold; *Ascda.* Pine Rivers

Left: *Kagawaara* Megawati Soekarnoputri

Above: *Ascda*. Nongkham x Muang Thong

Opposite, Clockwise from Top: *Ascovandoritis* Thai
Cherry; *Lewisara* Fatimah Alsagoff var. Zahrah
AM/OSSEA (1990); *Vasco*. Viboon Velvet

Singaporean hybridisers have pursued their unique line of breeding by bringing *Renanthera* and *Paraphalaenopsis* into combinations with *Ascocentrum* to make orchids that are sun-loving, tough, hardy, easy to cultivate and free-flowering, and the form need not be round. Two of these are perennial favourites: *Starmariaara* Noel (*Renanthopsis* Moon Walk x *Ascda*. Meda Arnold), registered in 1974, and *Bokchoonara* Khaw Bian Huat (*Arachnopsis* Eric Holttum x *Ascda*. Tan Chai Beng), registered in 1977. *Himoriara* (*Ascocentrum* x *Paraphalaenopsis* x *Rhynchostylis* x *Vanda*) will probably become just as popular. Other multi-generic combinations of the region are *Lewisara* (*Aerides* x *Arachnis* x *Ascocentrum* x *Vanda*), *Alphonsoara* (*Arachnis* x *Ascocentrum* x *Vanda*), *Duvereuxara* (*Ascocentrum* x *Paraphalaenopsis* x *Vanda*) and *Yusofara* (*Ascocentrum* x *Arachnis* x *Renanthera* x *Vanda*). In the 21st century, *Ascocendas* will not be bred in large numbers because excellence in their form, colour and size has largely been achieved. In Southeast Asia, they are now used as stepping stones to develop larger-sized, free-flowering, brilliantly coloured, heavy duty *Mokara* for the cut-flower market. An important milestone was marked by the awarding of the first-ever First Class Certificate by the Orchid Society of Southeast Asia (FCC/OSSEA) to *Mokara* Zaleha Alsagoff 'Nong' in 1997, in the 70th year of the society's existence and 39 years after it initiated its award system.

Genus: *Doritis*

Doritis is a unique genus with but a single species, *Doritis pulcherrima*, which occurs in diploid and tetraploid forms in the wild. For a long time, it was included in *Phalaenopsis* (under the name *Phalaenopsis esmeralda*), but it has been now definitely separated from the latter genus on the basis of its distinct floral structure. Its chromosomes are also larger than those of *Phalaenopsis*, and many hybrids (*Doritaenopsis*) between the two genera are sterile.

Doritis pulcherrima has a short, stout, erect stem up to 15 cm tall with stiff, lanceolate leaves 6–15 cm long and 2.5–5 cm wide. The leaves may be green, greyish green or purple, often with a purple flush on the underside. The floral coloration is unrelated to the leaf coloration. Numerous roots appear from practically every leaf axil and offshoots are freely produced at the base, each plant being a community rather than a solitary plant. The inflorescence is erect, up to a metre tall, carrying 12–20 flowers which open in succession. The inflorescence continues to elongate after the flowers have opened: on a strong plant, the fresh segments may keep flowering up to a year. The flowers face all directions and this characteristic is dominant in its first-generation intergeneric hybrids. Flowers are 1–4 cm across and possess a prominent lip which points vertically downwards like a spear. Hence the name *Doritis* from the Greek *doru* which means 'a spear'. The flowers are extremely variable in colour, from pure white, bicoloured (with a red lip), light pink to rose to a deep magenta. The deep purple forms have smaller, rounder, thicker and flat flowers, altogether better characteristics as a parent for breeding purposes.

The peloric *alba* form has ridges, spurs and yellow lip marking on the petals.

Flowers of the variety *buysonniana* are twice the size of the diploid *pulcherrima* but paler. The plants are also larger, with thicker leaves which are speckled with fine spots of purple on their upper surface. This variety was used to produce the larger, first-generation *Doritaenopsis*; when bred with tetraploid *Phalaenopsis*, fertile, amphidiploid progeny is produced.

Doritis pulcherrima is widely distributed throughout Myanmar, Thailand and Indochina to the northern half of Peninsular Malaysia and Sumatra. It is a lowland species growing in sandy soil along the coast in the shade of small shrubs or exposed on rocks near the sea. Some plants are deciduous. The variety *buysonniana* occurs only in northeastern Thailand near the banks of the Mekong in Ubol and its distribution does not overlap that of the diploid *pulcherrima*. No triploids have been reported among the wild species. The flowering season of *Doritis pulcherrima* is from June to December, with the variety *buysonniana* having a shorter season (from June to August).

Doritis pulcherrima is an easy orchid to grow and propagate. It is best grown in clay pots in charcoal and should be provided with good drainage. The plants root rapidly and extensively and soon produce *keikis* which can be removed to start new colonies or left behind to produce large clumps of plants. They can tolerate nearly full sunlight and may be liberally fed with either chemical or organic fertiliser.

Hybrids of *Doritis*

Up until 1960, *Doritis* was commonly classified under *Phalaenopsis*. Because of their close relationship, *Doritis* is more easily bred with *Phalaenopsis* than with any other genus. However, the chromosome sizes of *Doritis* and *Phalaenopsis* are different, and in the hybrid genus *Doritaenopsis*, there is poor chromosome pairing at meiosis, frequently leading to sterility. Another closely related genus is *Kingidium*, and with *Kingidium*

philippinensis the free-flowering *Doriella* Tiny was produced, vegetatively dominated by the *Doritis* habit. While flowering, it produces a free-branching inflorescence with flowers that closely resemble *Kingidium*. Primary hybrids have been produced by crossing with *Aerides*, *Ascocentrum*, *Neofinetia*, *Renanthera* and *Vandopsis*. Polygeneric hybrids are rare (*Ascovandoritis*, *Beardara*, *Hagerara*, *Hausermannara* and *Hugofreedara*).

The first *Doritaenopsis* was made in 1923 in Hawaii and registered as *Dtps.* Asahi (*Dor. pulcherrima* x *Phalaenopsis lindenii*). Then, for the next 36 years, no further hybrid was registered. In 1959, Clarence Schubert registered *Dtps.* Red Coral (a cross between *Dor. pulcherrima* and the pink, tetraploid

Opposite: *Doritis pulcherrima alba*

Above: *Dtps.* Queen Beer — an outstanding, large-flowered, first-generation *Doritaenopsis*.

Below: Typical look of a first-generation *Doritaenopsis*

Phalaenopsis Doris). It was an instant success and is one of the finest *Doritaenopsis* ever made, earning numerous awards including the First Class Certificate from the American Orchid Society. Charles Beard has bred the largest number of awarded *Doritaenopsis* during the 1960s and 1970s. Many of the wonderful Taiwanese *Doritaenopsis* are derived from the breeding trend that began in Florida.

In selecting a *Doritis* for breeding, colour is the most important consideration, followed by form. The tetraploid *Doritis pulcherimma* var. *buysonniana* is not always superior although it will produce fertile amphidiploid hybrids when bred to tetraploid *Phalaenopsis*.

Two types of second-generation *Doritaenopsis* have been bred:

1. By back-crossing to *Doritis* to increase the intensity of the colour.
2. By crossing forward with *Phalaenopsis* to produce larger flowers.

When the first-generation *Doritaenopsis* is bred back into *Dor. pulcherrima*, the resultant hybrids have upright sprays of sparkling crimson flowers which are well spaced on the spike, like a greatly improved *Doritis* (*Dtps.* Firecracker, *Dtps.* Fuschia Princess).

Although considerable difficulty may be encountered in carrying forward the *Doritaenopsis* into the second and subsequent generations, the effort is worthwhile because the results are often spectacular. *Dtps.* Ravenswood, *Dtps.* Maggie Fields, *Dtps.* Jane Sector, *Dtps.* Pueblo Jewel, *Dtps.* Clarelen, and *Dtps.* Elsa Radcliffe are examples of second-generation *Doritaenopsis* of the first type. The tetraploid *Phalaenopsis* Doris bloodline figures prominently in all these hybrids. The appearance of the *Doritaenopsis* up to the second generation is similar to the *Doritis* and the plants are easy to grow and propagate.

In the third generation along this line of breeding, the hybrids take on the appearance of *Phalaenopsis,* and in good clones the arrangement of the flowers is superb (as seen in *Dtps.* Kristine Teoh and *Dtps.* Jason Beard). Taiwan has carried forward the work of Charles Beard to produce the fine pink-red lip and striped *Doritaenopsis* of the fourth, fifth and even later generations (such as *Dtps.* Chia Ta, *Dtps.* Happy Valentine and *Dtps.* Yoshiko Beauty, etc.). Good semi-*albas* are now so commonly used for indoor landscaping there is considerable carelessness about labelling them.

Crossing with the colourful members of the *Stauroglottis* group and complex hybrids has introduced interesting colour forms (pure yellow, yellow with red lip, sunset shades, spotted pinks, yellows, peloric forms, etc.) Today, thanks to the work of the resolute Taiwanese breeders, the colour range and patterns of *Doritaenopsis* equal that of *Phalaenopsis*.

Modern, large pink *Doritaenopsis* are, for practical purposes, indistinguishable from excellent pink *Phalaenopsis* — but there are important differences in their behaviour. The *Doritaenopsis* are hardier and will readily bloom in the tropical lowlands, unlike pure *Phalaenopsis* hybrids of *Phal. schilleriana.* They also have an extended blooming period and the inflorescence is usually erect. The coloration is intense. Hence, many of the tropical pink *Phalaenopsis* on sale in the flower shops and gardens are, in fact, *Doritaenopsis*.

Opposite, Clockwise from Top Left: *Dtps.* Summer Red (*Dor. pulcherrima* x *Phal. cornu-cervi*), a member of the *Stauroglottis* group. Such breeding typically intensifies the colour of the *Doritis*; a complex intergeneric hybrid involving *Ascocentrum, Rhyncostylis, Vanda* and *Doritis*; *Dtps.* Ever Spring Prince (mutation); a typical modern Taiwanese semi-*alba Doritaenopsis* reflects the state of perfection which has caused the breeders to show no interest in naming it.

Left: *Dtps.* I-Hsin Golden Prince

Genus: *Gastrochilus*

This small genus of 15 species is distributed widely from Japan to the Himalayas to Java, with only two found in Malaysia. This genus is closely related to *Sarcochilus*. They are dwarf monopodial orchids with a few leathery leaves and produce short, simple inflorescences with a few relatively large flowers. The flowers are fleshy with a large, round sac to the lip.

Gastrochilus patinatus

This species from the Malaysian lowlands grows around limestone close to rivers. It has been found in Perak and Pahang and also occurs in Sumatra and Borneo. Its flowers are 2 cm across, yellow, lightly spotted with red.

No artificial hybrids have been recorded with the genus although the possibilities are fascinating.

Genus: *Kingidium*

Kingidium is very close to *Phalaenopsis* and has only recently been removed from the latter genus. The flowers are miniature *Phalaenopsis* with a lip that is larger than either the sepals or petals. The genus is widely distributed from southern India and Sri Lanka across Southeast Asia up to the Philippines.

Kingidium decumbens

This species is widely distributed and commonly found on old mangosteen trees (*Garcenia mangostana*). It is a small, *Phalaenopsis*-like plant with thin leaves 10 cm by 3 cm. The scape is thin, 6–12 cm long, with a few white flowers which face all directions. The lip is purple. There are fine purple spots at the base of the lateral sepals.

Kingidium philippinensis

This species grows in the lowlands. The flowers are 1 cm across, a pale yellow with purple on the lip. The inflorescence is erect and branched, often producing accessory branches for several months. This habit is seen in its hybrid *Kingidium* Tiny (with *Doritis pulcherrima*).

Genus: *Luisia*

This genus of long-stemmed, terete-leaved, sun-loving, monopodial orchids has some 40 members stretched across tropical Asia, from India and Sri Lanka to Myanmar, Thailand, Malaysia, Indonesia, the Philippines, Papua New Guinea and Polynesia. The inflorescence is extremely abbreviated but may carry up to 10 flowers that are typically greenish with a fleshy, purple lip that is distinctly divided by a groove into apical and basal parts. Holttum noted that the flowers continue to increase in size for a few days after opening, particularly in *L. jonesii* and *L. antennifera*. The hybrids of *L. jonesii* with *Vanda* are quite interesting. *Scottara* Memoria Andrew Leicester is a vigorous, floriferous hybrid between *Aerdns.* Bogor 'Apple Blossom' and *L. jonesii*.

Genus: *Neofinetia*

This genus has a single species that enjoys a wide distribution in temperate Asia, from China to Korea and Japan.

Neofinetia falcata

This is a small plant up to 15 cm in height ,with fleshy leaves that completely cover the stem. The inflorescence is short, with few white flowers that are characterised by narrow petals and sepals, a long, slim ovary and a filiform, curved spur that is up to 4 cm in length. It has been bred to *Aerides*, *Ascocentrum*, *Ascocenda*, *Phalaenopsis*, *Renanthera*, *Rhyncostylis*, *Vanda*, etc. Understandably, these hybrids were made principally in Hawaii, continental America and Japan.

On these two pages, from Left to Right: *Gastrochilus patinatus*; *Kingidium philippinensis*; A hybrid of *Luisia jonesii*; *Rumrillara* Sugar Baby (a hybrid with *Ascofinetia*)

Genus: *Phalaenopsis*

Phalaenopsis, in long arching sprays, are among the most beautiful flowers in the world. Their name is given for their resemblance to a flight of moths at twilight (*phaluna* in Greek meaning 'moth' and *opis* meaning 'resembling'). In the Philippines, they are called *mariposa* or butterflies. The White Moon Orchid, or *anggrek boelan* (*Phalaenopsis amabilis*), is the national flower of Indonesia.

Phalaenopsis are widely distributed throughout Southeast Asia, with a few species extending northwards to Taiwan and Sikkim, India and southwards to Australia and the Pacific. With only a few exceptions, they come from the lowlands and are epiphytes growing on trees close to streams. They are shade-loving plants requiring a light intensity of only 1,000 foot-candles and have responded well to fluorescent light culture and cultivation by the window sill. *Phalaenopsis* are excellent orchids for the beginner because they are simple to grow. The beautiful flowers last for 2–3 months, and near-continuous blooming can be achieved by good culture and judicious pruning of the flower spike. *Phalaenopsis* are widely cultivated in the lowland cities of the Asian tropics, but the most spectacular flowering is to be seen in France, Florida, California, Hawaii, Germany, England and Taiwan. The best hybridisation has been carried out in these areas.

Phalaenopsis are monopodial plants with large succulent leaves on an extremely short stem. They are capable of continuous growth throughout the year if they are kept in warm (above 20 degrees Centigrade), bright surroundings. In the temperate countries, growth slows down in autumn and ceases in winter. The peak flowering season is from December to February for the large, spray-type *Phalaenopsis*, while the small colourful species of the *Stauroglottis* group flower from April to September. The flower spikes arise from the basal parts of the stem and flower initiation is stimulated by a fall in night temperatures below 21 degrees Centigrade for a period of two weeks. Short days can also stimulate flowering. The flower spikes take 90 days to develop and each successive bud opens 3–4 days after the preceding flower. The flowers last up to three months and it is common to see large branching sprays with 15–30 flowers. When the stem has finished flowering, it can be pruned back to the first faded flower. And the node just below this will develop a side spray if the flower spike and the plant are strong. The flowers on the side spray open about three and a half months later, offering seasonal blooms in summer. These are usually smaller than the first blooms on the original spray.

In the Philippines, *Phalaenopsis* are grown in hanging pots or baskets, and it is a beautiful sight to see these plants with their cascading sprays strung along the verandah. Under such dry conditions, they must be watered at least twice a day. The preferred method is to grow them upright in plastic pots using any of a wide variety of potting media, from charcoal and brick to tree fern, redwood bark, osmunda and diverse potting mixtures. They also do well if they are tied to a tree fern slab or to a length of wood. Watering is varied according to the potting medium, air movement, light and temperature. They can tolerate temperatures between 21–32 degrees Centigrade in the daytime, but the night temperature should be between 15–22 degrees. They are heavy feeders and like an occasional application of organic fertiliser.

There are two major subdivisions in *Phalaenopsis*:

1. *Eu-phalaenopsis* which have large, round flowers, with the petals larger than sepals. The lips have fine appendages at the tip resembling antennae or horns. The members of this group have branching sprays of flowers well displayed as a cascade, the best examples being *Phal. amabilis* var. *grandiflora*, *Phal. schilleriana* and *Phal. sanderiana*.

2. *Stauroglottis* which have small colourful flowers, usually barred or spotted. Their petals are the same size or smaller than their sepals, and the lip is devoid of appendages; in some species, the lip is decorated with hair or calli. The characteristics of the lip are important criteria for the identification of the individual species within the group.

Phalaenopsis amabilis

All white *Phalaenopsis* are ultimately derived from this beautiful species which is widely distributed from Taiwan across Borneo, the Philippines, Indonesia, Papua New Guinea to Australia (Queensland). It was first described in 1750 by German botanist Rumphius, who found it on the island of Ambon (now called Sulawesi). Two years later, Peter Osbeck came upon this plant on the island of Teneli, west of Java, where only princesses were allowed to wear its blossoms. The name *Phalaenopsis* was give by Dr. C.C. Blume in 1825, because he mistook the flowers for a flock of butterflies when he first saw them through his field glasses.

Phal. amabilis is such an outstanding flower that those who came upon its various forms in the past have attempted to designate if by a new species name. It is now recognised that *Phal. grandiflora*, *Phal. aphrodite*, *Phal. formosana* and *Phal. rimestadiana* are all various forms of *Phal. amabilis*.

Phal. amabilis is a large plant with dark green leaves measuring up to 50 cm long and 10 cm wide. Its long, arching, frequently

Opposite: *Phal.* Sogo Lisa

branched sprays carry 12–24 pure white flowers which are 9–12 cm across. Suspended 30–40 cm above the plant, they indeed resemble a flock of white moths when seen at a distance. The throat of the flower is yellow with crimson markings. We have already described the *Phalaenopsis* flower in detail in Chapter 2, and it is sufficient to say that in *Phal. amabilis,* the petals are flat, round and larger than the sepals. Good forms of *Phal. amabilis* have flowers which are round with no gaps between the petals and the sepals. The variety *grandiflora* is extremely floriferous and may have 30–40 flowers on 4–5 side sprays. A very beautiful form is the variety *aphrodite* which has a greenish flush in the throat; it has smaller flowers and is found on the islands of Negros and Leyte in the Philippines, northwards to

Taiwan. Most white *Phalaenopsis* have a faint crimson colour on the undersurface of the petals and sepals, but the variety *rimestadiana* does not have this crimson flush and its blooms are of heavier substance than the other forms of *Phal. amabilis.*

Phalaenopsis sanderiana

This is a pink *Phalaenopsis* which was, at different times, thought to be either a variety of *Phal. amabilis* or a natural hybrid between *Phal. amabilis* and *Phal. schilleriana*. It is now regarded as a distinct species. Found at the southern part of Mindanao in the Philippines, it flowers in spring and summer. The flowers are similar to those of *Phal. schilleriana* but are paler, and the lip is different in shape. The

colour also fades when the plants are brought down to sea level.

Pink *Phalaenopsis* do not like high night temperatures and may refuse to bloom even after they have formed flower spikes; instead plantlets are produced where the flowers should have appeared.

Phalaenophsis schilleriana

This is a beautiful pink *Phalaenopsis* which is an essential component of all modern, large pink *Phalaenopsis*. It is a montane species found in the Tayabas province of Luzon and in the Laguna province of Mindanao (both in the Philippines), from 800–1,200 m above sea level. The plant has attractive mottled green foliage, which is purple on the underside, and flat roots which sometimes produce *keikis* at their tips. It flowers in February and March. The beautiful powder-pink flowers are fragrant, 9–10 cm across, borne on a tall, branched rachis, and they open in rapid succession. The lips vary from white to magenta. The sidelobes of the lips are flushed with yellow and spotted with dark crimson on its inner aspect.

To ensure flowering, growers in Manila send their plants up to the mountains in early winter to set the flower spikes. *Phal. schilleriana* is common on the outstretched branches of trees along the main highway into the mountains of Luzon, but they are only noticeable when they flower. When grown at sea level, the flowers are paler than when they are cultivated in the mountains, and this tendency to fade at warm temperatures has unfortunately carried into their progeny.

Phalaenopsis stuartiana

This species occurs at sea level in the northern part of Mindanao, yet it does not flower when cultivated in Singapore. Its leaves have a marbled appearance of green and grey-green on top while its undersurface is purple. The flowers are small, about 7 cm across, and rather open, but the inflorescence has many side sprays and usually carries 40–60 flowers. The flowers are white, with prominent cinnamon spots on the inner half of both lateral sepals and on the lip. *Phal. stuartiana* has bred some interesting freckled white and pink hybrids.

An unusual feature of this species is the appearance of plantlets along the longitudinal fissures of the root.

Subdivision: *Stauroglottis*

This group contains the small-flowered, star-shaped *Phalaenopsis* which come in a wide assortment of colours — green, yellow, brown, pink, rose, magenta and white. The sepals and petals are often barred or spotted, and the lip has no appendages. The lip is shaped like a spade, spear or anchor and may be decorated with hair and callosities. The characteristics of the lip are important criteria for their classification into species.

The species in this group are all very rewarding to grow as they are small and occupy little space. Several are extremely floriferous, yielding flowers throughout the year, but some only flower from summer to autumn.

Phalaenopsis amboinensis

Phal. amboinensis occurs in Sulawesi and the Moluccas. Three to five flowers of good form are carried on a rachis of 10 cm length. The two forms have basal colours of cream or golden yellow and both are heavily barred with cinnamon.

Because of the large numbers of fine forms available today, some outstanding spotted yellow hybrids have been made from it. The species is more free-flowering than *Phal. sumatrana, Phal. fuscata, Phal. lueddemanniana* and *Phal. gigantea* and is a good plant to grow. *Phal. amboinensis* has given rise to many fine yellow, spotted novelty *Phalaenopsis* when bred with other species and novelty hybrids.

Phalaenopsis bellina

Formerly known as the Bornean strain of *Phal. violacea* (the type strain comes from Perak in Peninsular Malaysia), this species is readily distinguished from the latter by its larger size, oval rather than square form and bowlegged sepals. The flowers, borne singly and up to three on a short rachis, are a pale jade-green overlaid with a magenta flush over the column, lip, lateral sepals and the base of the petals. The inner half of the lateral sepals is marked by a swath of deep magenta. The contrasting yellow of the lateral lobes of the lip provides a delightful contrast. Colour, form, size and shape are quite variable in the species. The flower is fragrant and attracts a species of wasp.

Phalaenopsis cochlearis

Phal. cochlearis is a pale yellow species from Borneo. It has an interesting striped lip. The hybrids have been disappointing.

Opposite: *Phal. amabilis*

Left: *Phal.* Taisuco Kaaladian exemplifies the modern white *Phalaenopsis* often profferred as a disposable pot plant.

Top: *Phal. stuartiana*

Phalaenopsis cornu-cervi

This is a small, star-shaped, yellow species which is very easy to grow into fine specimen plants. It is widely distributed throughout Myanmar, Thailand, Malaysia and Indonesia and has adapted to a wide range of light intensity and humidity. The plants growing in the shady habitats produce flowers which are superior in size, colour and form. The leaves are long, narrow and leathery. The characteristic feature is the rachis: it is branched and flattened distally with conspicuous bracts. *Keikis* are readily produced on old inflorescences.

Phal. cornu-cervi has many colour variants, and the size of the largest flower is three times that of the smaller-flowered forms. The usual colour is yellow, heavily barred with cinnamon.

Phalaenopsis equestris

Phal. equestris is a common lowland species from Luzon and other islands in the northern Philippines. Its branching inflorescence carries many small, star-shaped flowers with a prominent pink lip. The white petals are flushed with pink. The pleroic form is known as Star of Leyte.

All red-lipped *Phalaenopsis* have pedigrees that can be traced back to *Phal. intermedia*, the natural hybrids of *Phal. equestris*. It took many generations to produce bicoloured *Phalaenopsis* of moderate size. This is not well appreciated today because outstanding, large, bicoloured *Doritaenopsis* are common.

Phalaenopsis fasciata

Phal. fasciata is a yellow species from the Philippines commonly confused with *Phal. lueddemanniana* var. *ochreacea*. It is distinguished from the latter by its numerous backward-pointing callosities on the lip.

The flower is of a greenish yellow barred with brown, and the sepals and petals are curled backwards along its axis, more rounded at the tips and of a better substance than the *Phal. lueddemanniana*. The Fields strain (by Roy K. Fields) is in the background of many outstanding yellow and spotted hybrids.

Opposite, Left: *Phal. amboinensis*

Opposite, Right: *Phal. bellina* var. Phaik Khuan

Clockwise from Top Left: *Phal. equestris*; *Phal.* Be Glad
(a hybrid of *Phal. equestris*); *Phal. cornu-cervi* var. *alba*

Phalaenopsis fuscata

Phal. fuscata is found in shady forests by streams in Sumatra, Peninsular Malaysia and Borneo. It has few small flowers, 2.5 cm across; the distal halves of the tepals are yellow, the inner halves chocolate brown. The tepals tend to be rolled backwards and are cupped. The flowers open all at once, facing all directions. There is a natural hybrid between *Phal. fuscata* and *Phal. sumatrana*.

Phalaenopsis gersenii

Phal. gersenii is a natural hybrid between *Phal. violacea* and *Phal. sumatrana,* but it has more of the features of *Phal. violacea*. It is free-flowering.

Phalaenopsis gigantea

Phal. gigantea is named for its enormous leaves which measure up to a metre in length. It is found only in Borneo and is rapidly becoming extinct on the Malaysian side because of over-collection. It is not an easy plant to grow. The inflorescences are pendulous, hidden behind the enormous leaves, and carry 20–30 flowers which open all at once. It flowers from May to July and new blooms may appear on the extensions of the old inflorescence. The flowers are round, thick and waxy, drooping by their sheer weight. The tepals are a creamy white or yellow, heavily barred with reddish-brown. *Phal. gigantea* has produced many beautiful hybrids: these have an upright spike and the flowers are beautifully marked with stripes or heavy stippling.

The tendency towards large leaves is present in its first-generation novelty hybrids, but is not obvious when *Phal. gigantea* is bred to *Eu-phalaenopsis*. Hybrids of *Phal. gigantea* have flowers of heavy substance which are barred or spotted, the exception being the breeding to *Doritis pulcherrima* when the latter genus completely dominated the progeny. By and large, the first-generation crosses were not outstanding, but when carried forward, crosses with *Phal. gigantea* in their background have been quite outstanding.

Phalaenopsis lindenii

Phal. lindenii is a montane species from northern Luzon which has small white flowers that are spotted or striped with pink. The lip is oval, without callosities and is of a darker pink or striped. Neither the plant nor its hybrids will flower well in the tropical lowlands.

Phalaenopsis lueddemanniana

This lowland species which is found from sea level to 500 m is endemic all over the Philippines. It is a hardy plant with droopy leaves 6 cm across and 25–30 cm long. The 30 cm long inflorescence is sub-erect to pendulous and carries 1–3 star-shaped flowers 5 cm across. Their base colour is white to ivory, densely overlaid with transverse bars of magenta. The papillose midlobe of the lip is pink to crimson while the sidelobes are marked with a bright yellow over the keels. The plant is easy to grow and flowers in spring.

The species was described and named by the famous German botanist, Professor Heinrich G. Reichenbach, who also described three more varieties, var. *hieroglyphica*, var. *ochracea* and var. *pulchra,* on the basis of their markings and colours. Vegetatively, these varieties all look the same, but the variety *pulchra* flowers in summer, instead of spring.

In 1968, Herman R. Sweet separated four distinct species from the much confused

Phalaenopsis gigantea and its hybrids

Clockwise from Top Left: *Phal. gigantea*; *Phal.* Gretchen var. Gretchen HCC/AOS (*gigantea* x *stuartiana*); *Phal. gigantea* x *lueddemanniana*; *Phal. gigantea* x Hermione; *Phal.* Mok Choi Yew var. Phaik Khuan (x *violacea*).
(Photos of the hybrids with *Phal. stuartiana*, *Phal.* Hermione and *Phal. lueddemanniana* by courtesy of Dr. Henry Wallbrunn)

Opposite: *Phal.* Wanda Williams (*lindenii* x *amboinensis*)
(Photo by courtesy of Dr. Henry Wallbrunn)

Phal. lueddemanniana complex, using lip structure and other criteria to characterise the species. The interested reader is referred to his monograph, *The Revision of the Genus Phalaenopsis* published by the American Orchid Society in 1968.

Phalaenopsis hieroglyphica

This species was previously considered to be a variety of *Phal. lueddemanniana*, but since 1968 it has enjoyed species status. It is found in very shaded areas in the lowlands of the Philippines up to 500 m. The flowers are white to a light ochre overlaid with irregular transverse bars of cinnamon. The midlobe of the lip is white while the sidelobes are yellow.

Hybrids of *Phal. hieroglyphica* were listed under *Phal. lueddemanniana*. This 'variety' has not given outstanding hybrids when compared to the 'variety' *ochracea*.

Phalaenopsis manii

A species endemic in Assam, Sikkim and Vietnam, *Phal. manii* flowers from spring to autumn. The arching inflorescence carries 4–6 fragrant, yellow, waxy, star-shaped and bow-legged flowers that last a long time and thus appear as though they open together. The petals and sepals are commonly covered with faint cinnamon spots or are marked with light brown bars, while the anchor-shaped lip is white. The better forms have clear yellow flowers without spots and bars.

Its hybrids flower freely and remain in bloom for extended periods, but are lacking in fullness. Sweet observed that *Phal. boxalii* is indistinguishable from *Phal. manii*.

Phalaenopsis micholitzii

This rare Philippine lowland species has rather small, cream-white flowers that do not have any markings on the petals or sepals. There is a prominent yellow callus on the sidelobes of the white lip. The flower does not extend fully and is unimpressive, but Frederick L. Thornton started breeding with the species in the late 1960s. As far as the author is aware, only Michael Ooi of Penang, Malaysia has managed to produce anything worthwhile from *Phal. micholitzii*. This he achieved through a crossing with an awarded white *Phal. violacea* of the Perak strain, giving him a slightly larger, rounder, off-white *Phal.* Penang Violacea. Some clones are a clear yellow.

Phalaenopsis pantherina

This species is very similar to *Phal. cornu-cervi*, and in the past was considered to be one of the better forms of the latter. Peter O'Byrne says that it is found at the lightly shaded upper branches of tall trees in the lowland forests of Borneo. It has larger flowers and bolder barring, but the main distinguishing feature is the anchor-shaped midlobe of the lip which is of a bright white colour. It is distinguished from *Phal. manii* by the long isthmus to the midlobe of the lip.

Its alternative name is *Phal. luteola*. F.W. Burbidge who collected it at Labuan in November 1879 described his encounter in *The Gardens of the Sun* thus: "on some wet mossy rocks beside a rushing torrent, a glossy-leaved *Phalaenopsis* (*Phal. luteola*) displayed its golden blossoms, each petal and sepal mottled with cinnabar."

Phalaenopsis pulchra

The lowland species is endemic in the Philippines. The star-shaped flowers are very distinct, of a deep purple to magenta, and are devoid of markings. The dorsal aspect (back) of the sepals and petals are cream-coloured, and if one shines light through the flower, it is possible to distinguish bars on the petals and sepals. For this reason, it was once classified under *Phal. lueddemanniana*. But *Phal. pulchra* flowers in July and August, not in spring as most *Phal. lueddemanniana* do.

Phalaenopsis sumatrana

This lowland species enjoys a wide distribution almost throughout Southeast Asia, but it was first collected from Sumatra. It grows attached to tree trunks in well shaded areas near streams. The flowers are 5 cm across, with rather narrow but elegant petals and sepals, the latter set 45 degrees apart. The species is very variable. Sweet described five subtypes including *Phal. gersenii* which some people consider to be a natural hybrid with *Phal. violacea*.

Phal. sumatrana has been bred into several novelty hybrids and it is a possible source of yellow because its basal colour is a golden yellow.

Phalaenopsis venosa

This is a new species from Sulawesi, Indonesia which promises to be a good source of yellow, striped and spotted hybrids and perhaps some sunset shades. It is an epiphyte growing at an altitude of 700–1,500 m and produces flowers in spring. Many clones have rounded petals and sepals, with a greenish yellow base heavily overlaid with crimson bars that sometimes coalesce to form a solid reddish brown. The white column is surrounded by a flare of white at the base of the petals and sepals, and the midlobe of the lip is white as well.

There are already many awarded hybrids from *Phal. venosa*.

Opposite, Clockwise from Left: *Phal. lueddemanniana*; *Phal. pantherina*; *Phal.* pulchra
(Photo of *Phal.* pulchra by courtesy of David Lim)

Top: *Phal. sumatrana*

Above: *Phal. venosa*

Phalaenopsis violacea

A favourite species found in Perak in Peninsular Malaysia and in Sumatra, *Phal. violacea* produces star-shaped, pink, fragrant flowers throughout the year. The basic colour of the petals and sepals is a fresh apple green overlaid with a pink to maroon blush of varying intensity and extent. The midlobe of the lip is maroon and the sidelobes are a bright yellow. Some clones have a bluish tinge. An *alba* form is highly valued although some clones are less vigorous. While the flowers open one or two at a time, an inflorescence can continue to produce blooms for many months, and a strong plant will bear multiple inflorescences. Floral form improves as the plant gets stronger.

Taxonomists have now elevated the Bornean form of *Phal. violacea* as a separate species, named *Phal. bellina*. It is a daunting task to define the actual parentage of the hundreds of *Phalaenopsis* derived from *Phal. violacea*. A well-shaped *Phal. violacea* imparts an excellent form to its progeny for many generations.

Phalaenopsis viridis

Native to Sumatra and Peninsular Malaysia, this species is quite rare in cultivation. The sepals and petals have a greenish yellow base, overlaid with large, reddish-brown blotches. The spoon-shaped lip is white with purple striping on its ridges, much like *Phal. cochlearis*.

Hybrids of *Phalaenopsis*

Phalaenopsis is the most hybridised genus among the monopodial orchids with several thousand interspecific hybrids. Hybrid *Phalaenopsis* come in all colours of the rainbow except blue, and some are bicoloured, striped or speckled with spots. Intergeneric hybrids

Opposite Left: *Phal. violacea*, standard form

Opposite Right: *Phal. violacea* var. *alba* HCC/OSSEA (1982)

Left: *Phal.* (*bellina* x *violacea*)

Below: *Phal. viridis*

with *Doritis*, *Renanthera*, *Vanda*, *Ascocentrum*, *Neofinetia* and *Vandopsis* have added variety in floral form and coloration.

White *Phalaenopsis* were rapidly brought to perfection following the appearance of the tetraploid *Phal.* Doris in 1940 and the subsequent production of its numerous polyploid progeny. There are far too many early outstanding white hybrids for one to name them all, but the better known hybrids are *Phal.* Grace Palm, Elinor Shaffer, Ramona, Dos Pueblos, Bridesmaid, Joanna Megale, Alice Gloria, Cast Iron Monarch, Gertrude Beard, Daryl Beard, Jimmy Hall, Henriette Lecoufle, Miami Maid, Polar Bear and Wilma Hughes. Good hybrid white *Phalaenopsis* used to be classified into either: (1) exhibition type and breeding stock or (2) cut flower varieties, triploid hybrids being most suited for the latter purpose. This demarcation is now blurred, if not totally eliminated, through the input of Taiwanese breeders who have brought the commercial whites to perfection. Mericloned whites of award standard, in bloom, are now available for a song.

Whites were made more alluring by the addition of a dark-coloured lip, in red or in yellow with brown spots. The red-lipped hybrid has been produced by combining directly or indirectly with *Phal. equestris*, and sometimes with *Phal. violacea* and *Phal. lueddemanniana*. Initially, these semi-*albas* were smaller and of poorer substance than the whites, but a few awarded plants, for example, *Phal.* Arthur Freed 'Nuuanu', were magnificent. Further progress in this field was made by numerous growers working with the semi-*albas* produced by Arthur Freed Orchids. In addition, Taiwanese breeders used the semi-*alba Doritaenopsis* from Charles Beard (remember *Dtps.* Jason Beard). Several generations later, perfect semi-*albas* of pure *Phalaenopsis* and *Doritaenopsis* are almost indistinguishable from one another in terms of size, shape, substance, colour purity

and intensity and arrangement of blooms. One difference is that *Doritaenopsis* are hardier. From a tropical grower's perspective, the Taiwanese red-lipped *Doritaenopsis* are unbeatable.

The bicoloureds with a yellow brown lip have been introduced through crosses with *Phal. stuartiana*.

The breeding of pink *Phalaenopsis* took a great step forward with the discovery of a pink *Phal.* Doris (bred by Duke Farms of New Jersey), which paved the way for the appearance of tetraploid *Phal.* Zada (bred by Roy K. Fields in Miami) and *Phal.* Lavender Lady (bred by Herb Hager in California). *Phal.* Zada has in turn given rise to several outstanding pink hybrids. Pink *Phalaenopsis* breeding also benefited from the contribution of Marcel Lecoufle of France which also produced some beautiful hybrids. The problem with pinks is colour dilution: larger pinks are pale, dark pinks are small. This is due to the polyploidy of the white parent and the rarity of really dark forms of *Phal. sanderiana* and *Phal. schilleriana* from which many pink hybrids have been made; the third source of pink is *Phal. lueddemanniana* which is small-flowered. In the tropical lowlands, there is the additional problem of getting these plants to flower at the high temperatures; and when they do so, their colour is bleached because of the heat. The colour of pink hybrids is intensified and their substance is improved when they are crossed with members of the *Stauroglottis* section, but the resultant hybrids are few-flowered, small and star-shaped. When spotted pinks are crossed with coloured lips the resultant hybrid is candy-striped. Candy-striped plants have also been obtained by crossing semi-*alba Phalaenopsis* with members of the *Stauroglottis* section.

Here again, Taiwanese breeders have taken a different tack by breeding pink *Doritaenopsis* back to *Phalaenopsis* across several generations.

The F3 and later generations resulted in beautiful pink *Phalaenopsis* that are in reality *Doritaenopsis*. These plants are welcome in the tropics because they are hardy and flower readily. High temperatures do not pose a problem with flower initiation, although flower quality would be improved by lowering the temperature when the flowers unfurl.

These days, pink also comes with leopard spots.

The first truly remarkable yellow was Roy K. Fields' *Phal.* Golden Sands 'Canary' FCC/AOS, which had two sibs that won AM/AOS. It was registered as a cross between *Phal.* Fenton Davies Avant and *Phal. lueddemanniana* var. *ochracea*. However, Fields' *Phal. lueddemanniana* was actually a *Phal. fasciata* that was later crossed to *Phal. leuddemanniana* var. *hieroglyphica* to produce *Phal.* Spica. The latter plant also led to many generations of fine yellow hybrids (*Phal.* Barbara Moler and *Phal.* Golden Buddha and their progeny.)

The uniqueness of this particular *lueddemanniana* is that its progeny do not show premature fading. As Fields generously distributed his stud *lueddemannianna* to growers in the United States and in Taiwan (the author once owned a plant), today, one cannot be certain as to where the yellow is coming from. In 1969, Roy Fields showed me a plant of *Phal.* Goldiana whose flowers were covered with distinct spots of dark red throughout.

Opposite, Clockwise from Top Left: *Dtps.* Leopard Prince; *Phal.* Brother Lancer; A 10 cm *Doritaenopsis*; *Phal.* Sogo Grape

Overleaf, Clockwise from Top Left: *Phal.* Ever Spring King; *Dtps.* Nobby's Purple; *Phal.* Brother Romance

These pages continue to highlight the vast range of fabulous *Phalaenopsis* and *Doritaenopsis* hybrids from Taiwan.

Right: *Phal.* Ever Spring King

Bottom, Left: *Phal.* Ever Spring King x Taida Sweet

Bottom, Right: *Phal.* Golden Peoker x Taisuco Althea

Opposite: *Dtps.* Leopard Prince

Some of the spotted yellow hybrids we see today echo this line of breeding.

Fields found that *Phal.* Golden Sands 'Canary' FCC/AOS was not fertile, and it was a sibling plant variety 'Miami' AM/AOS that parented the many of 108 hybrids of Golden Sands that were produced between 1971 and 1991. Eventually, the Taiwanese hybridisers succeeded in breeding with 'Canary', but the initial hybrids were not superior to those made with the 'Miami' parent. The commonest yellow hybrid parent was Charles Beard's *Phal.* Barbara Moler (Donnie Brandt x Spica var. Florence) with 191 hybrids.

In Taiwan, *Phal.* Golden Sands was bred with *Phal. gigantea* to produce a fertile, spotted yellow named *Phal.* Liu Tuen-Shen that in turn produced beautiful progeny: bred to *Phal.* Freed's Danseuse, it enhanced the striping; to *Phal.* Percy Porter, it produced yellow with striping; to *Phal.* Spica, multiple spiking with beautiful star-shaped spotted flowers of exceptional substance; and in many cases, yellow with spots.

In the early days of breeding, *Phal. manii* was used an alternative source of yellow, but the plant did not produce overlapping flowers and the yellow colour of the F1 generation is pale. However, *Phal.* Bamboo Baby has been bred into the yellow line to give a clear yellow (without spotting). Many fine clones of *Phal. amboinensis* became available during the 1980s and they are behind many of the fine yellows we see today.

Over the past decade, *Phal. venosum* has been used in some breeding programmes, and it too has given some promising results, with many awards being granted to its hybrids by the American Orchid Society.

The *Phalaenopsis* which made the most waves at the end of the 20th century was *Phal.* Golden Peoker, itself a great hybrid and parent to a bewildering array of startling yellow and spotted *Phalaenopsis. Phal.* Golden Peoker

carries many species from the *Stauroglottis* section in its parentage, besides having *Phal.* Golden Sands, *Phal.* Barbara Moler, *Phal.* Bamboo Baby and *Phal. gigantea* as grandparents.

Another successful hybrid is *Phal.* Fortune Buddha. Of a dark yellow with spots, it has *Phal.* Golden Sands, *Phal.* Spica and *Phal. gigantea* as grandparents. It produced a beautiful spotted yellow when bred with *Phal.* Bamboo Baby.

Similar combinations account for much of the fine yellows, spots and striping that we have in *Phalaenopsis* today. Many exciting yellow hybrids and purple-blotched yellow hybrids appear in Taiwan every month.

Frederick L. Thornton and Dr. Henry Wallbrunn bred many novelty hybrids in Florida during the 1960s and 1980s. The trend was given an impetus when Arthur Freed's magnificent *Phal. violacea* imparted its full form and size to its hybrids. The appearance of many excellent clones of other *Phalaenopsis* species in the 1980s also assisted breeders to succeed in making many fine primary and secondary hybrids over the next 25 years. In Singapore, David Lim, with very modest setup, produced several superb crosses. Taiwanese hybridisers have collected many fine clones of the various *Phalaenopsis* species, and they could, if they wished, repeat the crosses to produce the ultimate first- and second-generation *Stauroglottis*-type *Phalaenopsis*.

Back-crossing to a plant featured in the background of a complex hybrid, especially back-crossing to a species, may highlight a desirable characteristic. Apart from producing the desired progeny, it discloses the breeding characteristics of that particular parent. Thus, the occasional step backwards may be a good way to move forwards.

Apart from the very successful breeding with the many varieties of *Doritis pulcherrima* and subsequent back-crossing to *Phalaenopsis*

to produce large flowers, *Phalaenopsis* has also produced fine hybrids when crossed with *Ascocentrum, Ascocenda, Renanthera* and *Sacrochilus*. However, the hybrids are difficult to look after and they did not inherit the tolerance for high light intensity from their second parent. Some of the well-known hybrids are 35–40 years old, and they reflect the special fortitude of their breeders who were willing to wait five years or more to see their hybrids bloom. Who can forget Iwanga's *Renanthopsis* Aurora and *Renanthopsis* Starfire (see p. 221), John Noa's *Renanthopsis* Ellen Noa or Thornton's *Asconopsis* Irene Dobkin? In 1995, *Renanthopsis* Carolina Sunset CCM/AOS served as a reminder of this special class. Unfortunately, such hybrids are seldom shown today.

Currently, Spots enjoy the limelight.

Below: *Phal.* (Auckland Buddha x Tailin Kaiulani)

Opposite, Clockwise from Top Left: *Phal.* (Fortune Buddha x Brother Goddess); *Dtps.* Chia Lin x (James Hall x Joanna); *Phal.* (Golden Grapes x *gigantea*) x *Phal.* Pinlong Cardinal

Opposite, Clockwise from Top Left: *Phal.* Little Emperor; *Phal.* Ching Her Buddha x Black Rose; *Asconopsis* Irene Dobkin; *Phal.* Salu Spot

Above: *Asconopsis* Irene Dobkin x *Phal.* Zuma Pixie

Genus: *Paraphalaenopsis*

Paraphalaenopsis is constituted by four species of a distinct group of rat-tailed (terete) *Phalaenopsis* which are endemic to western Borneo. The plants are really quite distinct from the typical flat-leaved *Phalaenopsis* and are all sun-loving, growing as epiphytes near riverbanks with their short stems hanging down and terete leaves arranged in a spiral fashion pointing downwards. They are characterised by a short scape and rachis which carry 5–15 flowers on each whorl and rather close together. The petals and sepals are similar — narrow, with an undulating edge and pointed at the tips — and about equal size. The sidelobes of the lip is erect and separated by a bilobed callus: the midlobe is widened at the apex. The apices of the lobes are a dark maroon while the base is white and spotted.

The three earlier species are distinctly coloured. *Paraphalaenopsis denevei* has 2.5 cm long tepals of a light warm brown fading to light green at the edges; *Paraphalaenopsis serpentilingua* has smaller white flowers, 3–4 cm across; *Paraphalaenopsis laycockii* is the largest and has pink petals and sepals 3.5 cm long. *Paraphalaenopsis labukensis* is coloured like *Paraphalaenopsis denevei*, but the flowers are heavily textured and 8 cm across.

Paraphalaenopsis denevei

This is the commonest and most popular species. The plant has only 4–6 stout terete leaves of a deep bluish green, each up to 0.5 m long. They flower from May to July. The rachis carries 3–15 flowers, each 5–6 cm across, with tepals of pale yellow-green to a golden yellowish brown and slightly faded at the border. The three lobes of the lip are solid crimson, while the centre is white with crimson stripes.

Previously, it was the custom to grow them in their natural fashion, with the stems hanging downwards, but nowadays it is common practice to grow them upright in

Opposite: *Paraphalaenopsis* Boedihardjo (*Paraphalaenopsis denevei* x *laycockii*)

Below: *Aeridopsis* Teoh Phaik Khuan (*Aer.* lawrenceae x *Paraphalaenopsis denevei*)

perforated clay pots in the manner of their hybrids which are always grown erect. If necessary, the leaves can be prevented from flopping by surrounding the plant halfway along its height with a circular ring of wire. They can also be grown tied to a tree, a stump of wood or tree fern.

Paraphalaenopsis laycockii

This is the rarest species. It is similar to *Phalaenopsis denevei* in most respects, but the flowers are larger, of poorer substance and of a light pinkish mauve. They were originally collected from central Borneo.

During the clearance of jungle for timber and the construction of massive dams in Borneo, large numbers of *Phalaenopsis laycockii* reached the market. This was followed a few years later by a large upsurge in the number of hybrids made from *Phalaenopsis laycockii*.

Clockwise from Top: *Renanthopsis* Dhanabalan. With the separation of *Paraphalaenopsis* into a new genus distinct from *Phalaenopsis*, the new name of the hybrid is *Pararenanthera* Dhanabalan; *Trevorara* Ursula Holttum 'Choo Kim Weng' HCC/OSSEA (1977); *Vandaenopsis* Catherine.
The terete *Vanda* are now also separated into a distinct genus known as *Papilionanthe*. However, for purposes of registration, they are still retained in *Vanda*. Hence, the new hybrid name becomes *Paravanda*, not *Parapapilionanthe*.

Opposite, Left: *Devereuxara* Ng So Peng

Opposite, Right: *Himoriara* How Xin Yi x *Vasco*. Five Friendship

Paraphalaenopsis labukensis

Little is known about this new species which is very difficult to flower in Singapore. It is the parent of the richly coloured *Vandaenopsis* Nelson Mandela.

Paraphalaenopsis serpentilingua

This delicate species has small ivory-white flowers which are accentuated by a lip that is alternately striped with yellow and crimson. The midlobe of the lip is forked like a snake's tongue at its tip, hence the name. The inflorescence is longer, with flowers more widely spaced than in the other species.

Hybrids of *Paraphalaenopsis*

The *Paraphalaenopsis* are interfertile with *Ascocentrum*, *Aerides*, *Arachnis*, *Vanda*, *Renanthera* and *Rhyncostylis*, but ironically, not with the *Eu-phalaenopsis* and the members of the *Stauroglottis* section of *Phalaenopsis*. *Paraphalaenopsis* produced beautiful primary hybrids when bred to *Aerides* (*Aeridopsis* or *Aeridoparaphalaenopsis* Teoh Phaik Khuan), strap-leaf *Vanda* (*Vandaenopsis* or *Paravanda* Prosperitas and Sophie) and *Renanthera* (*Renanthopsis* or *Pararenanthera* Amy Russell, Moon Walk, Taibar, Yee Peng and Dhanabalan). Its hybrids with *Arachnis* are not particularly attractive, but *Arachnopsis* (or *Pararachnis*) Eric Holttum, the first Singapore hybrid from a *Paraphalaenopsis* (*Arach.* Maggie Oei x *Paraphalaenopsis denevei*), registered in 1950, is an outstanding parent, and it has produced some extremely fine multi-generic hybrids (*Laycockara* Ian Trevor, Hong Trevor,

Sappanara Ahmad Zahab, *Trevorara* Ursula Holttum and *Bokchoonara* Khaw Bian Huat). These multi-generic hybrids have all been recognised by awards from the Orchid Society of Southeast Asia.

Several beautiful hybrids between *Vanda* and *Paraphalenopsis* were made after World War II, for instance, Lee Kim Hong's *Vandaenopsis* Prosperitas. The positive contributions of the *Paraphalaenopsis* to its hybrids are fine coloration (particularly with *Paraphalaenopsis denevei*), heavy substance, circular arrangement of its flowers and durability; while on the negative side, the inflorescence is stiff and

sometimes the scape is short, and the leaves have an unkempt appearance as they keep bending at the base.

Up till the end of the 20th century, *Paraphalaenopsis* were classified with *Phalaenopsis* in the Registration of Orchid Hybrids. Their recognition as a separate genus by the Orchid Committee of the Royal Horticultural Society today has created havoc with the nomenclature of the complex man-made genera containing *Paraphalaenopsis,* and there is much unhappiness in Southeast Asia over the loss of numerous generic names that evidently have priority. The registrants and

breeders of the following genera would like their generic names to be assigned to hybrid genera of *Paraphalaenopsis* and not be kept with *Phalaenopsis* as now proposed by the Orchid Committee:

Trevorara (*Arachnis* x *Paraphalaenopsis* x *Vanda*) (Ian Trevor, 1962); *Sappanara* (*Arachnis* x *Paraphalaenopsis* x *Renanthera*) (Syed Yusof Alsagoff, 1965); *Laycockara* (*Arachnis* x *Paraphalaenopsis* x *Vandopsis*) (Singapore Orchids, 1966); *Yapara* (*Paraphalaenopsis* x *Rhynchostylis* x *Vanda*) (Yap Kim Fatt, 1966); *Stamariaara*

Opposite, Left: *Vandaenopsis* Nelson Mandela x *Ascda*. Manili has inherited the rich coloration of *Paraphalaenopsis labukensis*.

Opposite, Right: *Paravanda* Joaquim's Child

Right: *Himoriara* How Xin Yi 'Veronica' HCC/OSSEA (1999) (*Paraphalaenopsis serpentilingua* x *Vasco*. Nong Kham)

Bottom: *Pararenanthera* — *Paraphalaenopsis laycockii* x *Ren*. (Kalsom x *philippinense*)

I have devoted several pages of this section to showcase the dedication of Singaporean and Malaysian breeders who explored the possibilities of *Paraphalaenopsis* and some of the fine results that they have achieved.

Overleaf, from Left to Right: *Bokchoonara* Khaw Bian Huat var. Khaw Phaik Suan AM/OSSEA (1977); *Stamariaara* APO (*Renanthopsis* Yee Peng x *Ascda*. Peggy Foo); *Yeepengara* Wong Yit Hoe (*Perreiraara* Luke Thai x *Paraphalaenopsis denevei*); *Wailaiara* (*Wlra*. Caroline x *Ascda*. Fuchs Gold)

(*Ascocentrum* x *Paraphalaenopsis* x *Renanthera* x *Vanda*) (Yap Kim Fatt/ Noel Sta Maria, 1974); *Edeara* (*Arachnis* x *Paraphalaenopsis* x *Renanthera* x *Vandopsis*) (Singapore Orchids, 1976); *Bokchoonara* (*Arachnis* x *Ascocentrum* x *Paraphalaenopsis* x *Vanda*) (Alsagoff/ Chan Sue Yin, 1977); *Himoriara* (*Ascocentrum* x *Paraphalaenopsis* x *Rhynchostylis* x *Vanda*) (How Wai Ron, 1997); *Yeepengara* (*Aerides* x *Paraphalaenopsis* x *Rhyncostylis* x *Vanda*) (How Yee Peng, 1997); *Waibengara* Wai Ron (*Aerides* x *Ascocentrum* x *Paraphalaenopsis* x *Rhynchostylis* x *Vanda*) (How Wai Ron, 2001)

Genus: *Pomatocalpa*

This is a genus of *Saracanthus* orchids with 30 members distributed from Sri Lanka to Samoa. The flowers are small, thick and fleshy and the shape is fairly consistent within the genus, with the lip pointing upwards, a wide sac-like spur behind and a narrow entrance to it. The name *Pomatocalpa* is derived from the Greek *poma* meaning a 'cover' or 'lip' and *kalpe*, a 'jar' or 'pitcher', referring to the shape of the labellum. Six species of *Pomatocalpa* occur in Malaysia. None is in cultivation and no hybrid has been made with the genus.

Pomatocalpa kunstleri

This is a beautiful dwarf plant with the stem only 10 cm tall, but it carries an erect, branched inflorescence of up to 30 cm length, bearing numerous small white flowers closely arranged all round the side sprays. It has a distribution covering Sumatra, Malaysia and Kalimantan, growing in the lowlands in shady locations.

It would appear that there is a tremendous potential for intergeneric hybridisation with *Pomatocalpa*, but the field is untouched.

Pomatocalpa spicatum

This is a small plant with thick leathery leaves up to 18 cm by 4 cm on a short stem. The inflorescence is single or with a few side branches, up to 15 cm long with many small yellow flowers closely arranged round the rachis. Many flowers, but not all, open at one time. The sepals and petals are marked with pink at the base.

It is distributed in Sumatra, Malaysia, Borneo and Java as an epiphyte in shady places.

Right: *Pomatocalpa kunstleri*

Below: *Pomatocalpa spicatum*

Above: *Ren.* Tom Story

Opposite: *Ren. philippinensis*

Genus: *Renanthera*

Renanthera are well represented in orchid collections in Southeast Asia because the red-flowered members are dazzling during their flowering season. There are a dozen species scattered from southern China downwards into Indochina, Thailand, Malaysia and the Philippines. Half of these are handsome and commonly cultivated, namely *Ren. storiei*, *Ren. coccinea*, *Ren. monachica*, *Ren. imschootiana*, *Ren. philippinensis* and *Ren. elongata*.

The flowers of most species are borne on large, horizontal inflorescences which are usually branched with 3–6 side sprays. The flowers are large or medium-sized, red and yellow, pure red or pure yellow. They have narrow petals and a narrow dorsal sepal,

but the lateral sepals are broader and drawn towards each other. The lip is small. The plants resemble *Arachnis,* to which they are closely related and with which they are interfertile.

The *Renanthera* may be grown in the same way as the Scorpion Orchids, either out in open beds in full sun or potted in brick and charcoal and supported by a stake. They need as much light as they can get in order to flower well. They also need to be watered twice a day and fertilised frequently.

Renanthera coccinea

This is the oldest species in cultivation in Singapore, where it flowers freely, almost throughout the year. In its native habitat, the peak flowering is from March to May. The tall, erect stem bears small leaves, 6 cm by 3 cm, of light yellowish green, and sends out branched inflorescences carrying 60–80 large flowers of Chinese red. A vigorous plant is so striking that many of its hybrids cannot match it in appearance.

Renanthera coccinea is a native of southern China, Myanmar, Thailand and Cambodia. The plants growing in Hawaii today originated from Singapore and were found to be hexaploid. Subsequent collections of plants from Thailand were subjected to chromosome analysis and found to be diploid, 2n = 38. There is also a tetraploid *Ren. coccinea.*

Renanthera elongata

Native to Malaysia, this species is found in the coastal lowlands in exposed localities and flowers well in Singapore. It is also distributed through Indonesia to the Philippines. The plant has dark green leaves and produces a complex inflorescence with secondary side branching on the side sprays. Its main drawback is that the flowers are tiny; 40–70 of them tightly cloistered on a 20 cm complex raceme with secondary side branches. Yellow clones are present.

Renanthera storiei

This is the premier species of *Renanthera,* widely used in hybridisation with more than a hundred hybrids to its credit. It is a native of the Philippines and is common in orchid gardens throughout the tropics. The plants are stout with thick, dark green leaves up to 20 cm long and 4 cm broad. The original strains with dark flowers were all tall-flowering, but a selfing in Singapore has produced short-flowering plants (flowering under 20 cm tall). The flowers are dark red with crimson blotches on the lateral sepals. In most plants, the lateral sepals have a wavy outline. The inflorescence is over a metre long, with side branches arranged in one plane, and may carry up to 150 blooms.

Renanthera philippinensis

This delightful species with evenly coloured flowers of good shape is found only in the Quezon province of Luzon, the Philippines, growing in mangrove. The inflorescence is 30–40 cm and carries up to 50 flowers.

Renanthera monachica

This beautiful orange-coloured, speckled species is a parent of the much admired early Hawaiian hybrid, *Ren*. Brookie Chandler. The species is usually described as few flowered, but Jim Cootes says it carries up to 50 flowers 4 cm across. The flower featured in his book, *The Orchids of the Philippines*, is perfect.

Other Species

Since the above five species are all free-flowering and produce very striking flowers, and because of the wealth of beautiful hybrid *Renanthera*, the remaining species are rarely cultivated: *Ren. imschootiana*, *Ren. matutina*, *Ren. pulchella* and *Ren. sarcanthoides*.

A new species, *Ren. bella*, produces brilliant red speckled flowers, 6 cm across, on a compact, unbranched spray. It should be interesting as a parent. It is found at the northern tip of Kalimantan, Indonesia, and is endangered from over-collecting. Of late, several new species have appeared. *Ren. augustifolia* is a cool-growing species which resembles *Ren. monachica*.

Leonid V. Averyyanov recently also reported two new species from the highlands of Vietnam abutting the Chinese border: a pale yellow *Ren. citrina* which is sparsely but attractively spotted with red, and a red *Ren. vietnamensis*. These plants are short and carry 20–25 flowers which appear to be related to *Ren. coccinea*. They are found at 700–800 m, in the open, on karst formations (coral beds pushed up from the sea bed that once covered the area). The area is rich in unusual

Opposite, Left: *Ren. augustifolia*

Opposite, Right: *Ren. storiei*. This is the superior clone that produces hybrids with no premature fading.

Right: *Aranthera* Beatrice Ng var. 'Conference Gold' AM/MOS (1963)

Below: *Aranthera* Gracia Lewis var. 'Burong Merah' AM/MOS (1964)

Both awarded clones of *Aranthera* were unique in their time as they did not show any premature fading of the early flowers when the spray was more than two-thirds open. The two earlier *Aranthera* in the cut-flower market, *Aranthera* Mohamed Haniff and *Aranthera* James Storie, both exhibited premature fading of the flowers.

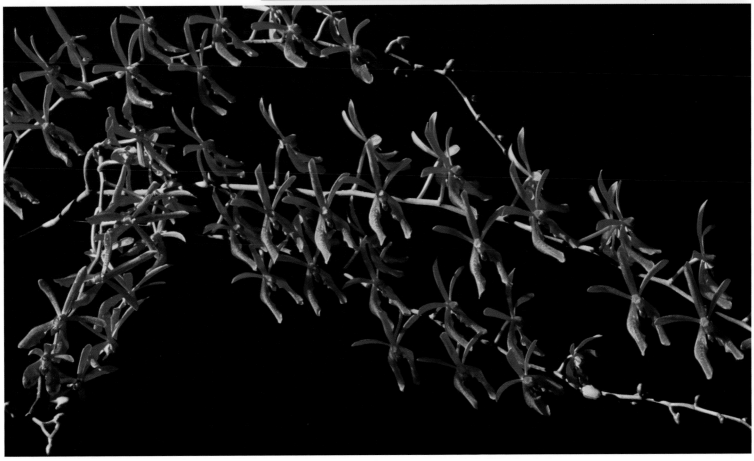

Red is predominant in hybrids of *Renanthera*. It may be modified if the other parent is tetraploid (as is the case with the *Renantanda*), or if one of the colour genes is missing, in which case the hybrid is yellow.

Clockwise from Top Left: *Renanstylis* Alsagoff; a Thai *Kagawaara*; *Renanthera* Yen

Opposite, Left: *Renanthera* Kalsom

Opposite, Right: *Renantanda* Carl Meier (*Vanda* Colourful x *Renanthera storiei*)

orchids, for instance, exciting 'Paphs' such as *Paph. malipoense*, *Paph. micranthum* and *Paph. henryanum*. Perhaps the Vietnamese *Renanthera* species are also present in Yunnan, China.

Hybrids of *Renanthera*

Interspecific hybrids within *Renanthera* have produced marked improvements on the species. *Ren. Kalsom*, *Ren. Kilawea* and *Ren. Brookie Chandler* are eye-catchers in any orchid collection.

The genus *Renanthera* has been bred with almost every single genus of horticultural merit in the *Vanda-Arachnis* tribe, although some combinations are difficult to achieve (such as with *Phalaenopsis* or *Doritis*). At the last count, there were 14 bi-generic, 32 tri-generic and 10 quadri-generic hybrid genera incorporating *Renanthera*. *Renanthera* is also a member of three hybrid-genera constituted by five genera.

Intergeneric breeding with *Renanthera* needs to achieve the following:

1. A red hybrid or one with saturated colours

2. Branching inflorescence
3. Good display of the flowers
4. Enhancement on the shape, substance and durability of red vandaceous orchids
5. Robustness and ease of cultivation

With the exception of substance, these goals are easily achieved. *Renanthera* imparts to its intergeneric hybrids its dominant red colour, floriferousness, a well-displayed inflorescence with many side branches, vigour and, in most instances, the ability to withstand full sun. In the rare instance when the dominant red overlay does not express itself (only in a few clones; most clones are red), the true basic yellow colour of *Renanthera* comes through, endowing the hybrid with an enchanting golden glow.

This happened with *Aranthera* Beatrice Ng 'Conference Gold' AM/MOS (1963), the yellow variety of *Aranthera* Gracia Lewis and *Renanthopsis* Aurora var. Gem AM/MOS (1964). The *Renanthopsis* Starfire var. Sunny AM/MOS (1963), a remarkable Hawaiian hybrid, carries a brick red blush over a yellow base. This area of hybridisation is largely ignored by current *Phalaenopsis* breeders, who might be deterred by the difficulty of raising such a hybrid and the long wait for it to bloom.

A much admired intergeneric hybrid is *Renanopsis* Lena Rowold (*Ren. storiei* x *Vandopsis lissochiloides*), a gigantic plant with an inflorescence to match. *Renanthoglossum* Red Delight also has a massive side-branching inflorescence with intense red flowers. Some quadri-generic hybrids, such as *Teohara* and *Andrewara,* did not show this branching habit.

Some of the prettiest hybrids are seen in crosses with *Phalaenopsis, Paraphalaenopsis* and *Vanda*. But though their flowers are large, they do not do well in a vase. *Paraphalaenopsis* imparts colour saturation, this being seen even in advanced hybrids. The miniature *Renanthopsis* Amy Russell and *Renanthopsis* Taibar are fascinating, with the terete *Phalaenopsis denevei* adding its impact on the shape and colour.

Premature fading is a common characteristic in hybrids with *Renanthera,* the red fading from the older flowers before the inflorescence has fully opened. It is most obvious in the *Aranthera, Renanthopsis* and *Renanstylis* but is now bred out in the modern hybrids which use superior clones of *Ren. storiei.* The earliest hybrids to overcome this weakness were *Aranthera* Bloodshot,

Arnth. Anne Black, *Arnth*. Gracia Lewis, followed by the tri-generic *Holttumara* Maggie Mason, *Sappanara* Ahmad Zanab, *Kagawaara* Teolone Fair, *Kagawaara* Boon Rubb and *Bokchoonara* Khaw Bian Huat. All received the Award of Merit. Awarded clones of *Renanstylis* Azimah, several *Renanstylis* Alsagoff, *Renanthera* Kalsom and *Renanthopsis* Dhanabalan are also free from premature fading.

A favourite crossing is between *Renanthera* and *Vanda* and almost a hundred such *Renantanda* have been made. Outstanding Singapore-bred *Renantanda* include *Rntda.* Prince Norodom Sihanouk, *Rntda.* Charlie Mason and *Rntda.* Ammani.

In the mid-1960s, several awards were given by the Malayan Orchid Society to *Renantanda* bred from the hexaploid *Ren. coccinea,* which is an even red, but this trend appears to have died out.

Improvement of substance and vase life appear to be dependent on the incorporation of *Arachnis. Arantheras* are the only hybrids that have been used in the commercial cut-flower industry. *Holttumara* is tough and long-lasting, but not free-flowering.

The commonest species used in all these hybrids is *Ren. storiei* (with over 120 primary hybrids), followed by *Ren. coccinea* (with about 40 hybrids). *Ren. storiei* has by far the best shape and size, with more flowers and numerous side branches, its drawback being that the flowers are spotted and not an even red. In this respect, *Ren. philippinensis* is better. Most hybridisers today still prefer to stick to *Renanthera storiei,* or else they would use a non-fading hybrid *Renanthera,* such as *Ren.* Kalsom, *Ren.* Tom Thumb, *Ren.* Yen or the diminutive *Ren.* Tom Story.

Right: *Renanthopsis* Starfire var. Sunny AM/MOS (1962)

Genus: *Rhynchostylis*

These beautiful Foxtail Orchids are the most striking orchids to have come out of Thailand, *Rhy. gigantea* itself being capable of producing no less than 1,000 flowers on 20 spikes on a single plant. In Thailand they are everywhere, and during the flowering season in January and February, the woods are filled with the fragrance from millions of blooms. According to Rapee Sagarik, in the mid-20th century, one could travel for days on end in northern Thailand without losing sight of flowering *Rhy. gigantea* on the trees. In their ideal natural habitat the plants are magnificent, each always bearing several clusters of superb white and purple flowers.

Rhynchostylis is a small genus with only four species, three of which are to be found in Thailand. They characteristically have medium-sized flowers which are tightly arranged in a cylindrical (Bottle-brush or Foxtail) form around the raceme. The inflorescence is erect in *Rhy. coelestis* and pendulous in the remaining three species. The plants grow on the main branches of deciduous trees and are exposed to strong sunlight when the trees shed their leaves during the dry season. The Thai species all require a decidedly low night temperature during bud development to flower properly. This presents a problem in Singapore and the plants are difficult to maintain, even in a vegetative state, although all their hybrids do extremely well. *Rhynchostylis* grows better in northern Peninsular Malaysia. In cooler latitudes, they are extremely easy to cultivate and flower, even in the lowlands. They do well attached to tree trunks or in open teak baskets. Good aeration and drainage is essential, and 50–60 percent shade is about optimum.

Rhynchostylis coelestis

This beautiful blue *Rhynchostylis* is distinguished from the other species by its upright flower spikes. It occurs in the mountainous regions of Thailand, except at Prachuab, where it grows at low elevations. It grows in deciduous forests, where it is subjected to a long dry season. The plant has the vegetative appearance of a strap-leaf *Vanda*, and each flower spike carries 50 fragrant light purplish blue flowers which are beautifully accentuated by a darker lip. The flowers are 1.5–2 cm across. There is also an *alba* form. The flowering season is from mid-April to July with peak flowering in May.

It is a rather delicate plant to grow because of its proneness to crown rot. Several attractive blue intergeneric hybrids have been made, such as *Rhynchovanda* Blue Angel.

Opposite: *Rhy. gigantea* var. 'White Elephant'
(Photo by courtesy of Professor Rapee Sagarik)

Right: *Rhy. coelestis* var. *alba*

Rhynchostylis gigantea

Rhy. gigantea, in particular the plum-coloured Sagarik strain, is the best known member of the genus. The flowers are of open form, 3–4.5 cm across and tightly arranged into an arching cylindrical inflorescence. The common variety has white flowers which are spotted with amethyst purple and accentuated by a purple blush on the lip. The size of the spotting varies, with the spots so dense in some that they coalesce to produce a solid, plum-coloured flower. Much of the attractiveness of *Rhy. gigantea* lies in its ability to produce three or more sprays of flowers at each flowering. The flowers are tightly clustered around the stem and produce a visual impact when they open simultaneously. The plants are seasonal bloomers, usually flowering only once a year from January to February.

Rhy. gigantea is widely distributed in northern Thailand above latitude 15 degrees north and from 15–1,500 m in various terrain. It also occurs in northern Myanmar and Indochina. It grows on the exposed branches of forest trees.

The requirements for prime flowering can be best appreciated by describing the climatic conditions. From May to October, the southwestern monsoon brings heavy rain, causing the night humidity to rise above 80 percent, although the day humidity may remain at 40 percent. The day temperature ranges from 8–30 degrees Centigrade depending on the altitude. November to February is the cold season when the night temperature may drop to 6–10 degrees Centigrade, or even 3 degrees Centigrade in the mountain valleys. This is the flowering season for *Rhy. gigantea*. The dry season sets in March and lasts through April, causing the trees to shed their leaves and leaving the *Rhynchostylis* exposed to more light and a wider fluctuation in day and night temperatures, in extreme cases as much as 39 degrees Centigrade in the daytime and

11 degrees Centigrade at night. But flowering is over with, and the plants now prepare themselves for vegetative growth by producing new roots.

Rhy. gigantea is miserable when it is grown in Singapore, and even if it should flower, the single inflorescence is puny, a ghost of the real plant in Thailand.

There are three exceptional varieties of *Rhy. gigantea*. The variety *illustris* has a similar coloration as the type, but the inflorescence carries more flowers with a narrower gap between the individual flowers, and the blooms are thicker and heavier. The variety *illustris* is found around Chiangmai at 160–880 m above sea level, and it can be distinguished by the stout stem, the leaves which are broader, shorter and thicker and the parallel green striping on the undersurface which is not as conspicuous as it is in the common variety. These plants, however, are not tetraploid (2n = 38).

The famous solid-red variety is universally known as the Sagarik strain, after Professor Rapee Sagarik, who, in 1954, first produced large numbers of the red clone by crossing two wild clones which had the deep colour. In the F1 generation, 80 percent turned up red and 20 percent were the common type, suggesting a single dominant gene with one homozygous parent. However, a previous selfing of the Ratana strain, the first red clone of *Rhy. gigantea* to be discovered, produced only the common variety in the F1 generation. Sagarik has also identified a tetraploid, plum-red variety of *Rhynchostylis*, which he has used extensively in his hybridising programme. The third outstanding variety is the *alba* form with spotless white flowers. It has been much sought after and has now been produced in quantity by selfing.

The influence of temperature in the flowering of the plum-red *Rhy. gigantea* is interesting. Flower spikes appear at the leaf axils from August to September towards the end of the rainy season, but they remain dormant until they receive a cold stimulus in December. If the temperature does not fall below 20 degrees Centigrade in December, or if the minimum night temperature of 12 degrees Centigrade lasts only a few days, the solid plum colour of the Sagarik strain will not develop properly. Colour breaks appear on the petals, particularly at the base, to reveal the white background. Conversely, if a white form is not the true albino, fine purple spotting may appear when low night temperatures are maintained throughout the period of flower development.

Rhy. gigantea, particularly the tetraploid Sagarik strain, has been used extensively in intergeneric hybridisation, producing several marvellous hybrids, such as *Renanstylis* Queen Emma, *Renanstylis* Alsagoff, *Rhynchovanda* Sagarik Wine, *Rhyncovanda* Colmarie and *Opsistylis* Lanna Thai. The breeding of the Sagarik strain *Rhyncostylis* to *Aerides lawrenceae* produced a remarkable hybrid bearing the traits of both parents.

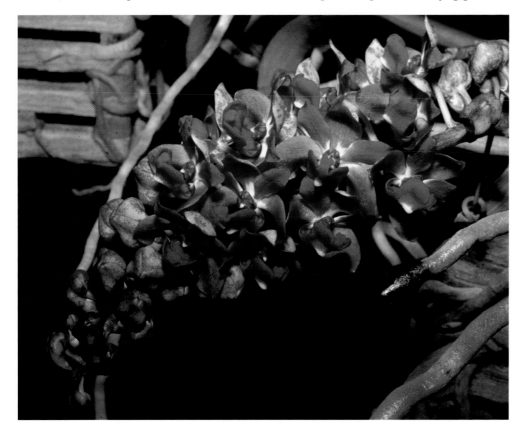

Opposite: *Rhy. gigantea*, typical form

Left: *Rhy. gigantea,* Sagarik strain

Rhynchostylis retusa

This is the common Foxtail Orchid which is widely distributed from Sri Lanka to the Philippines, across India, Myanmar, Thailand, Indochina and Malaysia. In Peninsular Malaysia, it occurs only in the northern states. Lowland plants are the best to grow, as mountain plants do not flower well when brought down to the tropical lowlands. Large plants are branched and do nicely when simply tied to a tree. The leaves are long and narrow (30 cm by 1 cm). The individual blooms are small, only 1.5–2 cm across, but about 100–140 flowers are closely packed around a drooping flower spike which is 15 cm long; a large plant can be extremely floriferous, bearing several dozen sprays which open simultaneously. The coloration is similar to *Rhy. gigantea*, but the amethyst spotting is minute. Again, the pure white forms are most striking. The flowering season is later, from April to May, with the flowers lasting for only a week.

The species does not grow well in Singapore but flowers readily in Penang and requires minimum care. It has not been used often in hybridising. The intraspecific crossings with *Rhy. gigantea* and *Rhy. coelestis* have both resulted in beautiful hybrids, and in each case, the spiking habit of the *Rhy. retusa* has been overshadowed.

Left: *Rhy. retusa*

Opposite, Top: *Neostylis* Luo Steary

Opposite, Bottom: *Renanstylis* Queen Emma

Rhynchostylis violacea

This rather uncommon *Rhynchostylis* is from Luzon in the Philippines. It resembles *Rhy. gigantea* but has fewer flowers. According to Holttum, this plant should grow more easily than *Rhy. gigantea* in Singapore, but it has not appeared in any collection on the island.

Hybrids of *Rhynchostylis*

The three species of Thai *Rhynchostylis* are all superb specimens, each with several colour varieties, and much effort has been directed towards the production of superior forms of the species through interclonal breeding. Professor Rapee Sagarik is the force behind Thailand's swift progress in orchid cultivation, and he has achieved singular success in producing the solid amethyst *Rhy. gigantea* which now bears his name (the Sagarik strain). The *alba* forms of *Rhy. gigantea* and *Rhy. coelestis* are also popular and have been bred in quantity; 'White Elephant' is the fanciful name given to the white *Rhy. gigantea*.

In the course of his work, Sagarik discovered a tetraploid amethyst-coloured *Rhy. gigantea* which has sired many fine intergeneric hybrids, among them *Rhychovanda* Sagarik Wine, *Rhyncovanda* Colmarie, *Opsistylis* Lanna Thai, *Rhyncovanda* Prinya Phornprapha and *Vasostylis* Malee Kanya. Although *Rhy. gigantea* does not flower well in Singapore, the pollen from its runted flowers is good and has been utilised to produce such outstanding crosses as *Renanstylis* Queen Emma, *Renanstylis* Azimah and *Renanstylis* Alsagoff, all of which flower continuously in Singapore. The *Rhynchostylis* form and colour are evident in the flowers of its hybrids, and in addition, it often produces gracefully arching sprays with closely arranged blooms. *Arachnostylis* Chorchalood (*Arachnis hookeriana* x *Rhy. gigantea*) has erect sprays with brilliant white flowers that are delicately spotted with purple.

Rhy. coelestis is an easier species to breed with, but it is a more delicate plant to grow, being commonly susceptible to crown rot. It has an erect inflorescence and has been used principally to produce dainty, blue miniature *Vanda* (*Rhychovanda* Blue Angel, *Rhychovanda* Blue Fairy and *Rhychovanda* Tan Gek Leng). Like the *Ascocenda*, these *Rhynchovanda* are free-flowering and have tall, erect inflorescences with a profusion of small flowers that face all directions.

The primary crossings with *Rhy. coelestis* give the deepest blue colour. Several intergeneric hybrids which retain the *coelestis* blue have also been made (*Yapara* Oi Yee, *Perreiraara* Porchina Blue, *Vascostylis* Tham Yuen Yee and *Neostylis* Luo Steary).

Genus: *Sarcanthus*

This is a large genus of 90 small-flowered, monopodial orchids which constitute an important part of the orchid flora in the Indo-Pacific region. The characteristic feature is the large callus at the back of the spur, just within the entrance. This spur is often filled with nectar. The flowers are fleshy and last several days. The name *sarcanthus* means 'fleshy', in reference to the texture of the flowers. The inflorescence is pendulous with a few flowers opening at one time.

Sixteen species are native to Malaysia, and they all occur in the lowlands. They are not commonly cultivated, and no hybridisation has been reported despite some interesting possibilities.

Genus: *Trichoglottis*

This is a moderate-sized genus of about 35 species which are distributed in tropical Asia, with a concentration in the Philippines. The name refers to the characteristic 'trident tongue', which is the midlobe of the lip. The stems are long, climbing or hanging, and bear few flowers (or a single flower) on short axillary inflorescences which develop simultaneously. The plants can be grown as for the *Vanda*. Several plants should be staked together to take advantage of the mass display of blooms during the flowering season.

With the exception of three species, *Trichoglottis philippinensis* var. *brachiata*, *T. fasciata* and *T. dawsoniana*, the members are not particularly attractive and are not commonly cultivated.

Trichoglottis dawsoniana

Unlike the other members, this species has a long, branched inflorescence with numerous flowers, 3 cm across. The narrow sepals and petals are like those of *Arachnis,* with a pale yellow base marked by transverse brown bars and blotches. It flowers in April and May. Although promising as a parent, no hybrid has been registered.

Trichoglottis fasciata

This species is found throughout Thailand, Indochina, the Philippines and northern Peninsular Malaysia, with a wide phyto-geographic distribution. In Peninsular Malaysia it has been found only in Kedah and Pulau Langkawi. It carries only 3–4 flowers on short stalks but flowers at practically every leaf axil; the flowers are fragrant, yellow striped with brown and have a prominent trident lip. It flowers in April and May.

Trichoglottis lanceolaria

This species has a hanging stem and is found on Gunung Panti and Pulau Tioman in Johore, Malaysia. It is also distributed in Java and Sumatra in Indonesia. The flowers are tiny, of pale yellow with a reddish brown band on the petals, and 2–4 flowers are present on each short spray.

Trichoglottis philippinensis **var. *brachiata***

This is the finest species in the genus, with beautiful flowers of crimson purple which are accentuated by a complex white lip. The flowers are fragrant, heavy-textured and last for several months from spring to summer. They are borne singly on each leaf axil.

Hybrids of *Trichoglottis*

Trichoglottis is not a favourite genus of the hybridisers since its single flowered inflorescence would severely limit the number of flowers produced by its first-generation hybrid. Nevertheless, the primary crossing between *T. brachiata* and *T. philippinensis* has resulted in a very striking hybrid whose ability to produce a long-lasting flower at every node more than makes up for the single-flowered inflorescence. Second-generation hybrids of *Trichoglottis* could be floriferous, and they have the additional assets of being colourful, heavy-textured and long-lasting. It would seem worthwhile to breed the more colourful, larger *Trichoglottis* with *Renanthera* and its related hybrid genera, such as *Renanopsis* and *Renanthoglossum*.

Opposite Top, from Left to Right: *Sarcanthus sp.*; *Trichoglottis brachiata*; *T. wenzelii* (syn. *T. geminata*)

Opposite Bottom: *T. philippinensis* var. *brachiata*

Right: *T. brachiata* x *T. philippinensis*

Genus: *Vanda*

Vanda is the Sanskrit name given to the orchids of Bengal. Today, it is the generic name of some 50 species which are native to tropical Asia. These are distributed from the Himalayas and southern China downwards across Sri Lanka and Southeast Asia to Papua New Guinea and northern Australia. *Vanda* are extremely popular, and they have been extensively hybridised among themselves and with other genera to produce offspring of exceptional beauty. A gift of 28 cuttings of Singapore's first hybrid, *Vanda* Miss Joaquim, to Lester Byron of Hilo initiated the orchid industry of Hawaii.

There are three distinct groups of *Vanda* which are distinguished by the appearance of their leaves, namely the strap-leaf, the terete and the semi-terete *Vanda*. The strap-leaf *Vanda* have flat leaves rather like a leather watch-strap, while the terete *Vanda* have cylindrical leaves like a pencil. The intermediate leaf form between the two occurs only twice in nature, but it is readily produced when a strap-leaf *Vanda* is crossed with a terete *Vanda*. Hybridisation has produced other horticultural types which are referred to as quarter-terete, three-quarter terete and so on.

Vanda are monopodial orchids. A well grown strap-leaf *Vanda* is a majestic specimen, tall, vertical with curved, fleshy leaves of equal size, neatly and regularly arranged in a fishbone pattern. Fat, cylindrical roots emerge from the stems perpendicular to the leaves and grow downwards and sideways, branching extensively when they reach a support. When the plant is tall, a cluster of plantlets arise at the base of the plant. Strap-leaf *Vanda* enjoy strong, dappled sunlight, an airy position, good drainage with regular feeding and

Opposite: *V.* Tan Hoon Siang

heavy watering. Thailand, which is the home of 12 indigenous species, is now the leading producer of many stunning hybrids. The Thais have developed a unique way of growing the strap-leaf *Vanda* in empty teak baskets: the roots of the plants hang downwards, reaching for the humidity of the soggy ground beneath the lath houses. In Bangkok, the immediacy of the Chao Praya River allows the grower to draw water into his land so that the ground never becomes dry. The method has caught on in Malaysia and Singapore, practically replacing all the earlier methods for housing these orchids.

Terete *Vanda* are tough, wiry plants which need as much light as one can possibly give them and must never be grown in the shade. The slender stems are sheathed by the base of their cylindrical leaves, and the entire plant is green except for the short internodes. The cylindrical leaves perform an important storage function, permitting the plants to withstand the dry season. Clinging roots enable the plant to hold on to trees and other objects for support. Feeding roots arise from the base of the stems, branching extensively when they reach ground humus or the soft mud of lowland swamps. Terete *Vanda* are grown in full sun, tied to wooden sticks and handled in the same way as the Scorpion Orchids.

The semi-terete *Vanda* inherit the vigour, toughness and floriferousness of their terete parents, and they are all sun-loving plants but do not grow as rapidly as the terete *Vanda*. The best way to grow them is in 15–25 cm clay pots in brick and charcoal and with heavy fertilising. In the commercial nurseries in Singapore, the growers leave a generous wad of matured pig manure permanently on the top of the potting medium for all semi-terete *Vanda*.

The inflorescences of *Vanda* are erect, usually unbranched, with large or medium-sized flowers of a wide colour range, in white, cream, yellow, orange, red, mauve, violet, blue and green and intermediate hues. Some flowers are tessellated while others are spotted. The sepals and petals are usually of equal size and round. The lip has three lobes and a spur. The overall flower structure is simple compared to most orchids.

The flowers of terete *Vanda* are frequently visited by large carpenter bees, and the natural hybrid *Vanda* Miss Joaquim was probably produced by insect pollination. Terete *Vanda* fade as soon as their pollinia are removed, and they release ethylene. In the confined space of a shipping box, the gas causes all the other flowers to fade, and this has made shipping difficult.

Vanda coerulea

This is one of the most handsome orchid species and is responsible for the entire range of beautiful round, blue *Vanda* and *Ascocenda* in cultivation. It is found at 1,000–1,500 m in northern and northwestern Thailand, extending across northern Myanmar to India. According to Professor Rapee Sagarik, who has made the finest collection of this species, including the pink form, the plants are usually attached to the trees in open forest and are well exposed to the sun. The climate of its natural habitat is cool and moist with night temperatures dropping to 10 degrees Centigrade. Plants of the wild species do not grow well in Singapore nor in the lowlands of Peninsular Malaysia, but they will flower in the hill resorts. However, new hybrid *V. coerulea* from Bangkok flowers well in Singapore. *Vanda coerulea* will thrive and flower well in temperate countries. The species is now endangered, but Thai breeders have produced such an excellent range of short-flowering, dark, tessellated, overlapping, round hybrid *V. coerulea* that few growers would

appreciate the jungle plant.

The inflorescence is erect and long, with 10–15 well arranged flowers, 10 cm across, in various shades of lavender to blue, rarely pink, with light or prominent tessellations on the petals and sepals, and a small dark blue lip. The petals are commonly twisted at the base, but hybrid *V. coerulea* now has excellent, overlapping, round form. *Vanda coerulea* is in the background of every blue *Vanda* and *Ascocenda*, and is also responsible for their tessellations.

Vanda dearei

This pure yellow *Vanda* comes from the lowlands of Borneo and flowers well in Singapore, where it has been used extensively in hybridisation. It is responsible for most of the large yellow *Vanda*. The short inflorescence carries only 3–6 flowers, which are fragrant. What it lacks in number it makes up for in size: the flower is 8 cm in diameter, in a full round form and of heavy substance.

Vanda denisoniana

Vanda denisoniana is native to the Arakan Mountains of Myanmar and in northern Thailand, where it is found at an elevation of 700–800 m. The plant flowers mainly in spring and carries a 15 cm inflorescence bearing 4–6 fragrant, waxy, extremely long-lasting flowers, 5 cm across, and of good form. The lip is white with a yellow blotch.

Left: *V.* Rose Davis (*V. coerulea* x *V. sanderiana*) x *V. coerulea*

Opposite, Clockwise from Top Left: *V. dearei; V. hindsii V. hastifera; V. denisoniana* x Luk Thai (two clones)

It was Sagarik who advocated the use of this species to produce in hybridisation because of its remarkably clear colours in the greenish yellow and orange range. Sagarik collected eight different colour forms of *V. denisoniana*, with two being more outstanding than the rest. *Vanda denisoniana* has resulted in a beautiful new range of *Vanda*, *Ascocenda*, *Aranda* and *Mokara* that are often fragrant.

Vanda hindsii

This new species is not on the hybridiser's list. It has medium-sized flowers, 3 cm across, of pale yellow, heavily marked with longitudinal stripes of brown.

Vanda hookeriana

A free-flowering *Vanda* with long, slender, branching, green stems bearing terete leaves 7–10 cm long and of 3–4 mm thickness, *V. hookeriana* is native to Borneo, Peninsular Malaysia and Sumatra, where it is found growing in swamps, open bush and belukar. It used to be present in abundance in the Kinta

Valley in Malaysia and was called the Kinta weed. The plant scrambles over bushes and small trees reaching for sunlight. The inflorescence is up to 20 cm long and bears 3–6 flowers of light mauve, 2–3 opening at a time. The petals are twisted at the base through 160 degrees. The lateral sepals are white and spread horizontally. The large lip is marked by a callus and is spotted with deep purple over the midlobe and flushed with deep purple over the side lobes. Quite an attractive flower by itself, it was made famous by its primary hybrid, the *Vanda* Miss Joaquim.

Vanda lamellata

This small-flowered *Vanda*, which is native to the Philippines, grows and flowers well in Singapore. It carries a tall, erect inflorescence 30 cm long which bears many light yellow flowers that are marked with brownish stripes. It was the first *Vanda* to be crossed with an *Arachnis*, a pioneering hybrid by Professor Eric Holttum, which he named *Aranda* Deborah.

Aranda Deborah was such a vigorous and free-flowering plant, it immediately inspired hybridisers in Singapore and Peninsular Malaysia to concentrate on breeding *Aranda*.

Vanda limbata

This is a medium-sized *Vanda* with waxy flowers, 3 cm across, of rather open form. The flowers have a yellow base overlaid with brownish blotches, and the lip is a light mauve. It is native to the northern Philippines and to Java, Indonesia. It grows well in the lowlands, in bright light, and flowers readily.

Vanda luzonica

This species, native to Luzon in the northern Philippines, is very similar to the Javanese *Vanda tricolor*. The fragrant flowers are white with crimson spots but some are pure white. It grows and flowers well in Singapore. The crimson spots are present in its hybrids.

Vanda merrillii

This is a robust, lowland Philippine species which branches readily and may reach a size of 1.5 m. When it is not in bloom, it looks very much like a *Vanda tricolour* with tough leathery leaves that are 3 cm wide and 30 cm long. The inflorescence is horizontal and bears up to 10 waxy, fragrant, very long-lasting, yellow flowers that are heavily marked with red but leaving a thin rim of yellow along the wavy outline of the sepals and petals. The large lip is trilobed, and yellow with red streaks. It flowers mostly in spring (April).

Its hybrids inherit the brilliant red colour and varnished appearance.

Vanda roeblingiana

Endemic to the Banguet and mountain provinces of Luzon, Philippines, at 1,600 m, it has flowered in the Singapore Botanic Garden's Cool House. The inflorescence carries up to

15 flowers, 6 cm tall and 5 cm across, facing all directions. They are yellow, heavily spotted with reddish brown. The fringed fan-shaped, brown-striped lip is most unusual and justifies some experimental breeding.

Vanda sanderiana

Vanda sanderiana is so spectacular, it is in a class of its own. For different reasons, taxonomists also propose that it be separated from the other *Vanda* species and given its own generic name, *Euanthe*. However, popular usage and the International Committee of Orchid Nomenclature still list *Vanda sanderiana* among the *Vanda*. This spectacular species from the Philippines was named by Professor Reichenbach for Frederick Sander, whose agent, Roebellin, first came upon the plant in 1880. It is the king of *Vanda* and is responsible for all the large round vandaceous hybrids.

The distinguished Malaysian orchidist, Datuk Dr. Yeoh Bok Choon loved to repeat the following doggerel, which highlighted the popularity of *Vanda sanderiana* among the hybridisers:

Little Vanda don't you cry;
You'll marry Mr. Sanderiana, by and by.

Vanda sanderiana grows on trees close to the sea on the southeastern part of the Philippine island of Mindanao, where it is known as the *waling-waling*, a name which means 'beautiful'. The plant has the habit of a robust strap-leaf *Vanda* and produces a 20 cm long inflorescence which generally carries 10–14 perfectly flat, round flowers closely clustered but well arranged around the stem. The dorsal sepals and petals are a milky colour with cinnamon brown blotches at the base, while the large lateral sepals are overlaid by a dense network of reddish-brown veins, some

On these two pages, from Left to Right: *V. hookeriana*; *V. lamellata*; *V. merrillii*; *V. luzonica*; *V. roeblingiana*

Above: *V. sanderiana*

On these two pages, from Left to Right: **A typical** *V. sanderiana* hybrid; *V. teres*; *V. tricolor*, typical form; *V. tricolor*, yellow strain

coalescing to form a large circular maroon patch. The lip is short with a hollowed base of pale green or dull yellow and is lightly streaked with red; the midlobe is a dull reddish brown. In its hybrids with other *Vanda*, four characteristics of the *sanderiana* persist: size, the full, round shape, the markings on the lateral sepals and the shape and coloration of the lip.

Intensive intraspecific hybridising in Hawaii during the 1950s and 1960s brought *Vanda sanderiana* to perfection.

Vanda scandens (*Vanda hastifera*)

A lowland species from Borneo and the Philippines, the species was named by Holttum for the scrambling habit of the plant, which sometimes reaches a size of 1.5 m. The upright inflorescence carries 3–8 fragrant flowers, up to 3 cm across, that are finely spotted with brown over the central half of the tepals.

Vanda hastifera is similar but has larger flowers.

Vanda spathulata

This is a yellow-coloured species which is native to Sri Lanka and southern India. The plant has the habit of an *Arachnis* with short leaves. It grows and flowers quite well in Singapore. It has small flowers on an erect inflorescence, but they open a few at a time. It is hexaploid and dominates every hybrid that it produces. The sole desirable characteristic is the pure yellow colour.

Vanda teres

Vanda teres is widely distributed in Laos, Thailand and Myanmar, reaching the foothills of the Himalayas. It is a climbing orchid usually found scrambling up tree trunks or bushes. The plants are more plump than those of *V. hookeriana,* and the flowers are rounder and larger, up to 10 cm across. The petals are twisted so that the back surface faces front. The colour is a delicate mauve. The variety *alba*

is exquisite; the variety *andersonii* has a striped, yellow lip. They all flower freely in Singapore.

Vanda tessellata

Another native of Sri Lanka and southern India, *V. tessellata* does not flower well in Singapore. The flowers are 5 cm wide and have a strong musk. There are many colour varieties, the commonest being yellow streaked with brown tessellations. Some flowers are clear yellow. Hybrids of *V. tessellata* inherit its tessellations and fragrance.

Vanda tricolor

The distribution of *Vanda tricolor* extends from Java to northern Australia. The inflorescence is curved, with 6–10 fragrant, waxy, long-lasting flowers, about 4 cm tall. The petals and sepals are curled backwards in its proximal two-thirds and widened in the distal third, with an undulating edge; the petals are also

twisted backwards. The colour is very variable, from off-white to pale cream or mauve, with numerous chocolate-brown spots arranged in longitudinal rows. Its hybrids display the mauve colour and spotting.

Much cultivated in Java before and just after World War II, *Vanda tricolor* is one of the parents of the tetraploid parent of 'The Pride of Hawaii', *Vanda* Nellie Morley.

Hybrids of *Vanda*

Hybrids of *Vanda* may be grouped into three different categories:

1. Interspecific hybrids within the same section, such as the strap-leaf hybrid *Vanda*.
2. Interspecific hybrids between members of two sections, such as the semi-terete *Vanda*.
3. Intergeneric hybrids

The interspecific hybrids within the same section are highly fertile, apart from those triploids which arise when one of the parent species is tetraploid. Singapore's famous *Vanda* Miss Joaquim, which launched the orchid flower industry of Singapore and Hawaii, is among the few natural hybrids which have risen under cultivation. The white *Vanda* Miss Joaquim var. John Laycock, which was bred by Datuk Dr. Yeoh Bok Choon from *alba* forms of *V. teres* and *V. hookeriana*, is charming and appears to be strongly influenced by the larger flowered *V. teres*. The other common *alba* hybrid, terete *Vanda* is *V.* Poepoe var. Diana. Crossed with a pale coloured *V. sanderiana*, it produced the beautiful white *V.* Lily Wong, which received AM/MOS in 1966. When these *alba* plants are grown exposed to full sunlight and rain, it is difficult to obtain spotless blooms during the wet season and, for cut-flower purposes, perhaps they should be grown under clear polyvinyl.

Vanda sanderiana and *V. coerulea* are probably the most outstanding *Vanda* species,

and it is no surprise that when they were combined to produce *V.* Rothschildiana, the result was stunning. Since the registration of the original cross in 1931, hundreds of repeat crosses have been made all over the world using superior parents, and today, extremely fine *V.* Rothschildiana can be bought at very reasonable prices from Thailand and Hawaii.

Strap-leaf *Vanda* are magnificent when they are in bloom, and there is a continuous interest in these hybrids. Using line breeding, a whole range of colours have been bred, with or without tessellations. Yellow comes from *V. dearei*, *V. denisoniana* (and *V. spathulata*, but this hexaploid plant is too dominant); orange from *V. insignis* and *V. denisoniana*;

On these pages, from Left to Right: Hilo Rainbow, a semi-terete *Vanda*; *V.* Velthius x Bill Sutton, a quarter-terete *Vanda*; *V.* Fuchs Delight 'Blue', a strap-leaf *Vanda*; *V.* Fuchs Delight 'Red', a strap-leaf *Vanda*. The *V. sanderiana* influence is evident in all four hybrids.

Both simple and complex vandaceous hybrids betray their *Vanda* parentage.

Opposite, Clockwise from Top Left: *Menziesara* Istana; *Vandaenopsis* Nelson Mandela x *Ascda.* Manili; *Mokara* Khaw Paik Suan x *V.* Kultana Gold; *Rhyncovanda* Colmarie

pink from *V. teres, V. luzonica, V. sanderiana, V. tricolor* and perhaps *V. coerulea*; red from *V. merrillii* and *V. tricolor*; and blue from *V. coerulea*; and fragrance comes from *V. denisoniana* and *V. tessellata*. The names of the famous crosses are a constant reminder of the tremendous effort put in by the Hawaiians (Ben Kodama, M. Miyamoto, T. Ogawa, Oscar Kirsch, John Noa and Roy Fukumura) and the Thais (Rapee Sagarik, Kasem Boonchoo, Thonglor, Phairot Lenavat and Nopporn Buranaraktham) to produce strap-leaf *Vanda* of superior quality: *Vanda* Hilo Blue, Hilo Rose, Mablemae Kamahele, OhuOhu, Bill Sutton, Frank Crook, Eisenhower, Ellen Noa, Waipuna, Onomea, Karen Koshiro, Jennie Hashimoto, Dawn Nishimura, Lenavat, Madame Chusri, Obha, Patou, Ratana, Amphai, Memoria Madame Pranerm, Thananchai, Pranerm Cloud 'Rung Ruang Ratana'. *V.* Robert's Delight and *V.* Fuchs Delight were bred by Robert Fuchs of Florida from Thai *Vandas*.

Vanda sanderiana improves the shape and size of the flowers in its hybrid, but the flowers tend to be clustered rather too tightly. *Vanda coerulea* extends the length of the rachis and improves the arrangement of the flowers. Its also imparts a beautiful blue coloration and tessellations. Roy Fukumura of Hawaii reckons that for a *Vanda* hybrid to be certain of having a good shape, there ought to be at least 65 percent of *V. sanderiana* in its make-up, assuming that ploidy is equal in all its parents. A polyploid parent exerts a tremendous

influence on its progeny, and this comment is then invalid.

The search for tetraploid strap-leaf *Vanda* is an important exercise: the chance discovery of only a single tetraploid strap-leaf *Vanda* (namely, *V.* Dawn Nishimura) has dramatically improved the range of cut flowers from Singapore.

The first amphidiploid *Vanda* were the semi-terete *V.* Emma van Deventer, raised by W. van Deventer in Java in 1926, and *V.* Josephine van Brero, raised by the eminent van Brero of Tjipaganti, Java, in 1936. Their achievements were remarkable because they had to sow the seeds among the roots of the pod parents, having no knowledge of asymbiotic culture at that time, and from among the single handful of seedlings, these wonderful amphidiploid plants were obtained.

Normally, intersectional hybrids are sterile because of the poor chromosome pairing at meiosis; but the doubling of chromosome sets in the amphidiploid clones restored full fertility. The appearance of *V.* Tan Chay Yan (the first cross between *V.* Josephine van Brero and a terete *Vanda*) in 1952 was a high point in Singapore's orchid history. It immediately earned the First Class Certificate from the Royal Horticultural Society and the Award of Merit from both the RHS and the Malayan Orchid Society.

It was the much needed shot in the arm for orchidists all over Southeast Asia, and over the next 15 years, a large number of *V.* Tan Chay Yan-type crosses were made: of these *V.* Tan Chin Tuan, Tan Yeow Cheng, TMA, and Patricia Low are outstanding. The more recent *V.* John Ede (*V.* JVB x *V.* Adrienne) maintains the high standard of the past. The apricot colouring of the amphidiploid *V.* Jospephine van Brero was transmitted to all its hybrids, even such intergeneric hybrids as *Renantanda* Ammani (x *Ren. storiei*),

Aeridovanda Aristocrat (x *Aerides lawrenceae*), *Christieara* Malibu Gold (x *Aeridocentrum* Luk Nok) and *Ascocenda* Karen Codling (x *Ascda.* Mem. Choo Laikeun), as well as to secondary crosses with *V.* Tan Chay Yan and TMA. It took a tetraploid blue *V.* Dawn Nishimura to produce a 'blue' quarter-terete *V.* Chia Kay Heng. These quarter-terete *Vanda* are extremely floriferous, being triploid, and severely subfertile. However, a few secondary hybrids of note have been raised, such as *V.* Sek Hong Choon (*V.* Tan Chay Yan x *V.* Ernest Fuginaga) and *Rntda.* Charlie Mason (*V.* TMA x *Ren. storiei*). *Vanda* Tan Hoon Siang is a recent outstanding hybrid of the JVB type which commemorates the name of the late president of the Malayan Orchid Society, who led the team that brought the Fourth World Orchid Conference to Singapore. It is worthy of note that AOS judges awarded a First Class Certificate to *V.* Nonthaburi 'Giant' (JVB x *V.* Gulf of Siam) in 1998.

'The Pride of Hawaii' is *V.* Nellie Morley FCC/AOS, the cross between the tetraploid *V.* Emma van Deventer and *V. sanderiana*. It is also triploid, but it is less vigorous than *V.* Tan Chay Yan. Only one hybrid has been registered with *V.* Nellie Morley as parent. *Vanda* Nellie Morley var. Nita and *V.* Tan Chay Yan var. Subha have been mericloned and are available at remarkably low prices.

It is very rare for an orchid to produce unreduced gametes. In the case of the diploid semi-terete *V.* Merv. L. Velthius, the chromosomes of the gametes were not only unreduced, W.B. Storey found that they were reduplicated, giving rise to sex cells with 4 sets or 76 chromosomes. When crossed with *V. coerulea*, it produced the pentaploid (5n) *V.* Nora Potter. It is important to remember that not all amphidiploid semi-terete *Vanda* make good parents. *Vanda* B.P. Mok produced large numbers of deep-coloured hybrids but all were affected by bud drop.

Vanda has been bred with every worthwhile genus in the *Vanda-Arachnis* tribe. Whenever the two genera are closely related, as in the case of the strap-leaf *Ascocenda*, the resultant hybrid is fertile, unless it is triploid.

Among the bi-generic vandaceous hybrids, *Aranda* is the most widely cultivated, dominating the orchid flower industry in Singapore and Malaysia. Whatever the species of *Vanda* used as parent, the hybrid *Aranda* always withstood the full equatorial sun. The miniature *Vanda* produced by crossing *Vanda* with such diminutive genera as *Asocentrum*, *Rhynchostylis*, *Aerides* and *Neofinetia*, come in all the colours of the rainbow, and they have earned a permanent place in every orchid collection. They are also extremely hardy plants which can be grown even in empty wooden baskets. The larger flowers of the second and later generations of *Ascocendas* have given rise to a new class of Spider Orchids (the *Mokara*) which boasts a new range of colours in the yellow-orange and red end of the spectrum.

Directions for Pure *Vanda* Breeding

Vanda Miss Joaquim var. Agnes is the oldest *Vanda* hybrid. Several attempts were made to improve on it, and one outcome is the delicate var. Josephine registered by John Laycock in 1938. Such terete *Vanda* are deficient in the areas of form, colour range, substance and flower life. Breeding to strap-leaf *Vanda* produced improvement in many areas, but the ideal quarter-teretes are still to come. A grouping of modern strap-leaf *Vanda* is pictured here. They illustrate the ability of *V. sanderiana* to impart a perfectly round form and size. However, the spacing of the flowers is poor; they are not naturally flat; and colour is not always bright. In *Vanda* breeding, the improvements should come in these deficient areas. Pure *Vanda* may give way

to *Ascocenda* because this direction of breeding continues to produce bright colours, excellent shape, improved arrangement, floriferousness, ease of cultivation and flowers which are indistinguishable from pure *Vanda*.

Opposite, Top: *Vanda* Miss Joaquim var. Josephine

Opposite, Bottom: *Ascda*. Tubtim Velvet x Kuchan bears a strong imprint of *V. sanderiana*.

Left: A group of strap-leaf *Vanda*. These also bear a strong imprint of *V. sanderiana*.

Genus: *Vandopsis*

This genus is closely related to *Arachnis* and *Trichoglottis* and is distributed from the Ryukyu Islands in Japan southwards through China, Indochina, Thailand, Malaysia, the Philippines and Indonesia to Papua New Guinea. With the exception of *Vandopsis parishii* var. *mariottiana*, which is unusually beautiful, the members of the genus are such large plants that they would usually not appeal to growers because they bloom infrequently and their flowers are not particularly attractive. However, the species *Vandopsis lissochiloides* has attracted considerable attention because of the outstanding hybrids that it has produced, among them the gigantic, red *Renanopsis* Lena Rowold and a host of giant Scorpion Orchids. The species which appear most promising today are *Vandopsis parishii* from Thailand and *Vandopsis warocqueana* from Papua New Guinea.

Vandopsis are robust plants, some being similar to giant strap-leaf *Vanda*, while others resemble *Arachnis*. They can have the same treatment as *Vanda* but must receive full sun and frequent applications of fertiliser.

Vandopsis gigantea

This species is distributed around Myanmar, Thailand and northern Malaysia (at the Langkawi islands off the western coast and in the islands off Trengganu on the eastern coast). It has also been found in the southern state of Johore. It is lithophytic, growing close to the sea. The plant is huge, with thick leaves of a pale yellow-green measuring 35 cm in length. The 35 cm long inflorescence carries 15 blooms which are round and dull yellow with reddish brown blotches. It flowers seasonally and irregularly.

Vandopsis lissochiloides

This species is found in the Philippines, Thailand and the Moluccas. It grows tall and must be firmly staked in cultivation. It flowers well in Singapore. The inflorescence is 1.5 cm long and carries 25 or more flowers, but they open three or four at a time. The substance is extremely heavy, and each flower is about 6 cm across, of greenish yellow spotted with reddish brown. The petals and sepals are bright purple or yellow on the back. It can flower continuously throughout the year, peaking in summer.

Vandopsis parishii

Unlike the preceding two species, this plant has a short stem with broad, fleshy leaves 20 cm long and 7 cm wide, with an appearance more like *Phalaenopsis*. It flowers seasonally from March to May, producing a horizontal inflorescence of 30 cm length with 6–12 flowers in two rows which last a fortnight. The flowers are 6 cm across, and in the variety *mariottiana*, it is round with petals and sepals overlapping and of extremely heavy substance. The tepals are brownish purple fading to light lavender inwards and white at the base; they are heavily spotted with dark purple.

This species is a native of deciduous forests in the mountainous northern regions of Myanmar, Thailand and Indochina and does not thrive in the tropical lowlands. It is a promising parent and has already produced some beautiful hybrids, such as *Opsistylis* Lanna Thai.

Vandopsis warocqueana

This lithophytic species are giant plants growing on rocks and cliff surfaces in the lowland areas of Papua New Guinea and its neighbouring islands. The stems are robust, 1–3 m long, densely leafed with leathery, tongue-shaped leaves about 35 cm long. The plant bears several inflorescences simultaneously, each 20–30 cm long, with side sprays bearing many flowers, each 2.5 cm in diameter. The basal colour is cream to yellow, heavily spotted with dull red, while on the reverse side, the petals are a dark purple. The flowers are long-lasting.

Mrs. Andree Millar has found a colour variety which is epiphytic on trees in Papua New Guinea. The plant is short and is very promising as a parent, particularly for crossings with *Arachnis* and *Renanthera*.

Hybrids of *Vandopsis*

The first hybrid between *Vandopsis lissochiloides* and *Renanthera* was stupendous, and nothing that has been made with *Vandopsis* has measured up to this *Renanopsis* Lena Rowold. The original cross was awarded throughout the world, receiving First Class Certificates from both the American Orchid Society and the Royal Horticultural Society. But these were tall flowering plants, allowing one to examine its flowers only if one stood on a very tall ladder. In the late 1960s, repeat crosses made in Singapore with short-flowering *Ren. storiei* brought the plants down to a more manageable height. And at least in South Asia, *Renanopsis* Lena Rowold flowers at a height of 0.5–1 m. The sprays are 1.5 m long with side-branching, the flowers are of extremely heavy substance, a bold red with crimson spotting, and the better clones show no sign of premature fading. The back-cross to *Ren. storiei* (*Renanopsis* Cape Sable) is also fine but uncommon.

Renanopsis Lena Rowold was crossed with *Aranda* Kian Kee to produce the tetra-generic *Teohara* Teoh Cheng Swee, which was named after the author's father. Side-branching was lost in the process. With *Phalaenopsis* it has

Opposite: *Renanopsis* Lena Rowold

Right: *Vandopsis gigantea*

Far Right: *Vandopsis warocqueana*

produced a hybrid which is similar to the bi-generic *Renanthopsis*.

Several giant, hybrid Scorpion Orchids with inflorescences reaching over 2 m were produced by crossing *Vandopsis* lissochiloides with various *Arachnis* (i.e. *Vandachnis*). The flowers were spaced far apart and were seasonal, and the plant occasionally required shock treatment to get it to bloom. The usual shock treatment is decapitation (better known as top-cutting, i.e. removing the top of the stem). After the mid-1960s, the interest shifted towards the more compact tri-generic hybrids, such as *Laycockara* (*Vandopsis* x *Arachnopsis*), which flower throughout the year. Both *Laycockara* Ian Trevor, which has yellow flowers, and *Laycockara* Lee Kim Hong, which has brick red flowers, are popular.

Opsistylis Lanna Thai, a cross between *Vandopsis parishii* and *Rhynchostylis gigantea*, is also beautiful, the good clones benefiting from the influence of the tetraploid *Rhynchostylis gigantea* used by Professor Sagarik. This is a small plant, quite unlike the enormous *Opsistylis* derived from *Vandopsis gigantea* which is hardly worth growing. Giant-sized orchid plants are decidedly out of favour, and any breeding involving *Vandopsis* must be able to produce compact free-flowering plants.

The Leafless Orchids

Beneath the velamen, many orchid roots are green and contain photosynthetic cells which are capable of supplying the plant with carbohydrates. This capability is most strongly manifest in the Tapeworm Orchids, *Taeniophyllum*, a large genus comprising some 180 species which are distributed from Japan and northern India to Australia. *Taeniophyllum* are members of the *Vanda-Arachnis* tribe, but they have become so modified that the leaves are reduced to tiny, brownish scales which cover a short apex-stem. All photosynthetic activity is performed by the flattened roots which are perpetually green. There are 17 species of *Taeniophyllum* in Malaysia, but they grow on the small branches of trees and are difficult to spot. The flowers open for only a day, in night-flowering species for only a night, some species flowering gregariously.

From Bangkok, we have *Chiloschista usnoides,* which produce 10–15 cm long, pendulous racemes, each bearing 10–120 beautiful round, lime-sherbet green flowers that last several weeks from March to May. The plants are difficult to maintain in cultivation, possibly because they are dependent on their mycorrhiza for organic compounds. These orchids must be retained on their original tree branches, and growers should try omitting the usual fungicidal programme considered essential for other orchids. Alex Hawkes suggests that leafless orchids should be tied on to a living tree and never be allowed to dry out.

A difficult category are the saprophytic leafless orchids, such as *Galeola*, *Gastrodia* and *Epipogum*, found in Asia and Australia, which belong to a different subfamily, *Neottiodeae.* They grow in primitive forest and are completely dependent on fungus for their food supply, not at all easy to find and nearly impossible to keep alive for very long in greenhouse conditions. As far as I am aware, they are not cultivated.

Opposite: *Hawaiiara* Firebrand

Right: *Chilochista usnoides*

Chapter 6

Requirements of the Orchid Plant

The basic requirements of plants are warmth, water, light, aeration, anchorage and nutrients. In order to grow orchids well, it is essential to understand how their individual needs are met by their structural adaptations (Chapter 2) to their natural habitat (Chapters 4 and 5).

If you are just starting to grow orchids, it is a good idea to visit a few orchid nurseries in your neighbourhood to see how they are grown. Ask the experienced nurseryman for his suggestions as to which plants would be suitable for cultivation indoors or outdoors. Everything depends, of course, on where you are living. For a start, it is important to select those orchids which are hardy, quick to grow and easy to flower. After you have grown some plants for a few months, you will get to know their preferences and their dislikes. Orchid plants are quite expressive, and if you have been treating them well, they will respond by producing handsome flowers, successively larger leaves and stronger roots which crawl all over the pot. If your handling is wrong, the plant remains stagnant. The root tips may dry up or become soggy and rotten, the leaves may lose their lustre and become shrivelled — bottom leaves turn yellow and wilt, new leaves are smaller. If there is insufficient light, the plant turns a dark green and never flowers.

Temperature

Orchids are classified into three categories on the basis of their temperature requirements.

Left: *Vandas* grown in the open at the Mandai Orchid Gardens in Singapore

The Cool-growing Orchids

These orchids grow best at night temperatures of 10–13 degrees Centigrade and day temperatures of 16–21 degrees Centigrade. They come from the mountainous regions of the tropics, for example, the large-flowering *Cymbidum, Odontoglossum, Miltonia, Sophronitis* and some *Paphiopedilum*; else, they are natural species of the temperate zone, such as the *Cypripedium* and the *Phragmapedilum*. In the cooler parts of the United States and in Europe, they are the most economical orchids to grow since they can be grown in a simple cool greenhouse which requires minimal heating. These plants have feeble growth in warm climates and never flower. Some attempts have been made to hybridise them with warm-growing orchids (for instance, interspecific breeding in the *Cymbidium* and *Odontoglossum* tribes) to produce hybrids which are more tolerant of higher temperatures.

The Intermediate Group

This group embraces the broadest collection of cultivated orchids. These plants prefer night temperatures of 10–16 degrees Centigrade and day temperatures of 21–29 degrees Centigrade. *Cattleya, Oncidium*, nobile *Dendrobium*, some *Paphiopedilum, Epidendrum* and *Miltonia* and a whole range of interesting botanicals belong to this group. Several orchids which usually grow in a warmer climate, such as *Phalaenopsis, Ascocentrum* and strap-leaf *Vanda*, will also thrive in the warmer parts of the intermediate orchid house.

If the growing temperature for an orchid

belonging to this group exceeds the ideal for the particular plant, when it blooms, the flowers will be fewer, smaller and of poorer shape.

The Warm-growing Orchids

These require night temperatures which are consistently above 16 degrees Centigrade for proper growth. They are all natives of the tropical region, particularly the tropical lowlands. All members of the monopodial *Vanda-Arachnis* (*Aerides*) tribe, including *Phalaenopsis* (apart from a few species), belong to this group. The horn *Dendrobium* (*Ceratobium* group), the famous Cooktown Orchid (*Dendrobium biggibum*), the Jewel Orchids, the mottled-leafed *Paphiopedilum*, the giant *Grammatophyllum* and many interesting botanicals such as the *Eria* and the *Bulbophyllum* also belong to this group.

It is actually a heterogenous grouping, and in Singapore, we further separate them into warm-growing orchids which require partial shade (night temperatures 16–22 degrees Centigrade, day temperatures 27–32 degrees) and warm-growing orchids which require full sun, tolerating day temperatures well above 38 degrees Centigrade and usual night temperatures of 20–24 degrees. The Scorpion Orchids, terete *Vanda*, terete *Phalaenopsis* and *Renanthera* require full sunlight for optimum growth and flowering. Intermediate *Dendrobium* grow and flower better with a bit of shade (about 25 percent). But what is surprising is that, once upon a time, *Vanda* Miss Joaquim, the famous Singapore hybrid, actually flowered better in England than its country of origin! Warm-growing orchids prefer cooler nights when they are in bloom.

While hybridisation between the intermediate and warm-growing species have resulted in hybrids with a wider temperature tolerance, by and large, the preference of the hybrid would be for the lower temperature range.

In the tropics, it is the simplest thing to grow orchids out in the open sun, and shade-loving types thrive under a lath house. A cool house is required for the intermediate group in the lowlands. In the highlands, they can also be cultivated in a lath house. The cool-growing types cannot be grown in the tropical lowlands.

In the temperate region, one has a choice of (1) erecting a special glass house where the suitable growing temperature can be maintained both in summer and winter or (2) growing orchids in a modest way by the window sill or under fluorescent light. In practice, most growers construct an intermediate house, and some may try to adapt a few cool-growing types and a number of warm-growing types to this intermediate environment. It is also possible to partition the orchid house into a warm-growing section and an intermediate area, or to produce a gradation of warm to intermediate by putting in a water-cooling system at one end of the plant house so that the coolest area is nearest this cooling system. If the only option is to grow orchids within the house, then one must choose those plants which require low intensity light and intermediate temperature, such as *Paphiophedilum* and *Phalaenopsis*.

Water and Humidity

Orchids are well adapted to dry conditions and are often none the worse after several days without water. In their natural environment, they can tolerate drought or just as easily withstand weeks of heavy rainfall. However, many orchids survive under conditions which are far from ideal. It is a common observation that when they are removed from the wild and grown in the greenhouse, where they receive frequent watering and application of fertiliser,

they are lush and vigorous in their growth and spectacular in their flowering. Strong, turgid leaves permit greater transpiration with improved transport of solutes, enhanced gaseous interchange and photosynthesis.

The frequency with which one should water the plants depends, among other things, on:

1. The way the plant is grown: pot or slab, type of pot, size of pot, drainage provisions in the pot, potting media, etc.
2. Atmospheric temperatures
3. Light intensity
4. Air movement
5. The condition of the plant
6. Whether the plant is epiphytic or terrestrial
7. The species of the plant

As a rule, atmospheric humidity of 40–50 percent is ideal for orchid cultivation in temperate countries; a humidity of 70–80 percent is preferred for orchids grown under shade in tropical countries; orchids grown in full sun may require a humidity in excess of 80 percent. Many commercial nurseries in Malaysia and Singapore, following the example of the Thai growers, keep the ground in their lath houses soaking wet to promote rapid growth of seedlings. However, orchid leaves do not absorb water from the atmosphere if the humidity is less than 98.8 percent, unless the plant is suffering from great water deficit, during which it can absorb water from the atmosphere at 92 percent humidity.

For the grower with small and medium-seized collections, the simplest way to promote higher humidity and reduce the need for frequent watering is to grow the plants in plastic pots which are positioned above a trough of gravel or to place foliage plants under the bench on which the orchids grow.

A few orchids have been grown successfully by hydroponics and in the Luwasa system, but this requires a larger capital investment and is really more complex. Instead of watering by hand, a sprinkler system can be installed and set to go off either at a set time of the day or when the humidity reaches a critical level. In commercial nurseries, these devices are not frequently used as it takes a worker very little time to hose down all the plants in the greenhouse or garden.

In the confined space of a glass house, orchids need not be watered more than once every 3–4 days in a cool climate, but in summer and in tropical areas, it is necessary to water once a day. When growing orchids in the tropics, you should water them once a day if you are using plastic pots and twice a day if the plants are grown in clay pots or on slabs of tree fern. When orchids are grown outdoors in the tropics, they have to be watered twice a day, morning and evening, unless there is rain.

There are two ways to water the plants: drench them until they are dripping wet or spray a fine mist along the rows of plants so that only the leaves and the surface roots are wet. If the plants are grown in the open and there is strong air movement around them, they should be drenched at every watering. For epiphytes grown under shelter in a lath house or in a greenhouse, most growers will use fine spray to economise on water usage, fertiliser and chemicals. The orchid roots and its contact surfaces soak up water within a few seconds, and continued watering merely causes the excess to run off. Some growers advocate a heavy drenching once or twice a week to leach off excess chemicals which have accumulated. However, for more than 50 years orchid growers in Singapore have grown the *Vanda* Miss Joaquim on stumps of wood on which are applied several layers of cattle manure; the master grower Emile Galistan even advocated dipping an entire wooden slab,

its covering osmunda fibre and the roots of the attached *Phalaenopsis* plants into a paste of cattle manure to promote the growth of the plant. Nowadays, commercial nurserymen in these parts place a very liberal application of pig or chicken manure on potted *Vanda* and *Aranda* to promote rapid growth and maximum flowering.

Drenching is necessary for terrestrials grown in soil because the roots are deep within the ground and spraying does not reach more than just the surface layer unless it is prolonged.

Light

In nature, most epiphytic orchids grow on the branches of tall trees and receive only dappled sunlight through the forest canopy. Such orchids require 25–50 percent shade in summer if they are cultivated in temperate countries, and throughout the year if they are grown in the tropics. The *Dendrobium*, *Cattleya*, *Phalaenopsis* and strap-leaf *Vanda* belong to this group.

In the Indo-Pacific Basin, some members of the *Arachnis-Vanda* (or *Aerides*) tribe grow along the seashore and in the exposed parts of the lowlands where they receive the full blast of the sun throughout the year. The only part of the plant which may be shaded are the feeding roots, which are hidden in the undergrowth. These orchids and their hybrids like full sun throughout the year and will only flower if they receive strong sunlight. Cover for their feeding roots is optional. Several commercial growers in Singapore still follow the old practice of heaping cut lallang (*Imperata cylindrica*), grass or wood shaving about the feeding roots, and many farmers probably achieve a similar effect (perhaps unintentionally) by slapping on a thick layer of pig manure on top of the pots. Most hobbyists and a few nurserymen grow these orchids with their roots exposed.

In choosing a site for growing your orchids, we suggest you select a plot which faces east, which benefits from full morning sunlight. The reason is not so much the quality of light but the temperature. A plot which faces west is warmer and the plants are more likely to suffer from heat shock and water stress. Flat dwellers, in particular, should choose a balcony or window sill that faces east to grow their orchids. Seasonal changes in the source of sunlight make it far more difficult to cope in a growing space that faces north or south, while a west-facing area will definitely require some shading and a setup to cool the plants.

The terrestrial orchids have a much lower light tolerance for they grow either in the deep shade of the forest, in bogs or in open scrubland where they are protected from the full sun by other plants growing in the bush. Under cultivation, these require 60–75 percent shade, or about 10,000–15,000 foot-candle hours per day. Examples are *Paphiopedilum*, *Cypripedium* and the Jewel Orchids. Other terrestrials, such as *Arundina graminifolia*, *Cymbidium finlaysonianum* and *Spathoglottis plicata*, which are among the early colonisers reclaiming land laid waste by natural catastrophes, thrive in open country. In cultivation, such plants should be given full sunlight or at most very light shade. The best *Arundina* that we have seen growing in the wild grow on steep hill slopes at 1,000 m, facing east (where they are in shade by mid-afternoon.)

The type of light is also important for proper growth of the orchid. Light in the ultra-violet, near ultra-violet and green ranges represses plant growth, and green PVC sheets are totally unsuited for roofing of orchid houses. When too much green algae collect on top of the plastic roofing, it similarly represses plant growth. The algae must be scrubbed off or the roofing replaced. The simplest approach is to employ shading which will not interfere

The orchids on these two pages illustrate the range of light requirements.

Left: *Paph. haynaldianum* (top) and *Calanthe veratrifolia* (bottom) both require very little light (about 20 percent daylight) and will do well in deep shade. However, they may be grown amidst orchids in up to 40 percent light.

Above: *Den.* Burana Jade x Srimahapo. The near *Den. phalaenopsis* hybrid is best grown in a lath house with 40–50 percent shade, but it may tolerate up to 75 percent sunlight.

Opposite, Left: *Mkra.* Citi Gold (top). The *Mokara* are usually grown in full sun for best flowering, but they may receive 10 percent shade; *Lc.* Chunyeah 'Good Life' GM/DOG (bottom). The *Cattleya* tolerates strong sunlight. This plant has been grown in full sun for the past year. It blooms readily, but the large flowers do not live up to the shape expected of this prize-winning plant when it blooms in Taiwan.

Opposite, Right: *Grammatophyllum scriptum* var. Yunnan Tiger. *G. scriptum* prefers a sunny position, but unlike its tougher cousin *G. speciosum*, it should be shielded from strong afternoon sunlight. Some growers in Bangkok have free-flowering specimen plants growing in 25 percent shade in the company of *Cattleyas*.

with the normal spectrum of sunlight, such as with lath houses, white paint over glass, or black saran cloth. For lath houses, the ratio of the width of the slats to the space between slats determines the amount of light which passes through. Since light in the morning and late evening is of a lower intensity, some growers consider it worthwhile to allow more of morning and evening light to get through and less of noonday light: this is achieved by having two rows of slats, one above the other, so that more light passes through in the morning and evenings. The strips of wood should run in a north-south direction to permit intermittent light and shade to fall on any point in the lath house throughout the day.

For vegetative growth, circadian rhythms are unimportant, and orchids can be grown in continuous light day and night to achieve maximum growth. However, day length, or more correctly night length, affects flowering.

According to Dr. William Standord, the amount of natural light received (intensity x day length) by orchids under cultivation in temperate zones is probably inadequate for proper photosynthesis and flowering. This is often also true for plants grown in many collections in the tropics. Increased light, higher humidity and lower day temperature would almost always lead to better growth and flowering.

Aeration

Epiphytic orchids cannot be potted in soil because their roots require good aeration. If they are waterlogged, they soon begin to rot. As the plants grow and are repotted into successively larger pots, bigger pieces of crock, charcoal, fir bark or tree fern have to be added to ensure that water is drained off effectively after it has wet the roots. After two or three years, the root ball becomes too extensive and the potting medium would have deteriorated

and collapsed, causing a constriction of the air spaces which then become waterlogged. Lichen, bacteria and even moss appear on the surface of the potting medium, and from then on, the plants deteriorate. To avoid this, most growers repot at least once in two years. An alternative approach is to top-cut the monopodial plant or to remove the forward bulbs from the sympodial orchids, leaving the back-shoots and back-bulbs to produce *keikis,* which can be removed and repotted whenever they produce sufficient roots.

In Southeast Asia, *Vanda, Ascocenda, Rhynchostylis* and *Aerides* are commonly grown in empty teak baskets to provide maximum aeration for the roots and total drainage after watering. These epiphytes like air movement — but they should not be hung to dry. If you choose to plant your orchids in baskets, there are two ways to avoid their drying out: by allowing a layer of water to remain below the orchids (this is exactly what they do in Thailand), or by placing shade-loving foliage plants under the benches (which is what nurserymen do in Singapore). If you do not adopt either system, your plants in baskets will be too dry. The roots develop rapidly if the plants are watered at least twice a day and fertilised frequently. In a few months such plants develop a magnificent branching root system; but this method is very wasteful of fertiliser.

Commercial growers in the tropics pay a lot of attention to wind movement when selecting a site for an orchid nursery. Strong wind movement is favourable, and a good site should have sloping terrain and should not be obstructed by tall structures. Orchid leaves are heated by radiant energy and can only lose this heat by conduction into the atmosphere. It is therefore important to have good air movement and high humidity around the plants since water conducts heat more efficiently than dry air. If the plants are

watered in the evening, a strong air current will also lower night temperature and induce flowering. Some growers also believe that drying the leaves and crown by air movement prevents bacterial and fungal rot.

Anchorage

In their natural state, all orchids grow as epiphytes on trees, or as lithophytes clinging on to rocks, or as terrestrials in the ground. Epiphytic orchids do extremely well if they are planted on living trees and given enough light and wind movement. But they will do equally well when planted in clay pots, perforated clay pots, baskets, a piece of crock or a stump of wood or charcoal. In Thailand, many vandaceous orchids are grown in empty baskets, and their roots trail below the baskets without making any attempt to seek attachment. Thus, it would seem that root anchorage is unnecessary for these plants. Nevertheless, the plants are firmly tied to the basket. Sympodial orchids require root attachment, and if the roots of the orchid plant do not make an attempt to anchor itself to the intended support, it would indicate that there is something wrong with the plant or the intended support.

Nutrients

There is no doubt that orchids benefit from the application of fertiliser, and it is difficult to believe that popular opinion once maintained that it was unnecessary to feed the plants at all or that it was even harmful. Dr. Davidson's explanation of the reasons behind this misconception is to the point, namely, "that orchids are no different from other plants in their requirements but they may take a longer time to show up their deficiencies because of their slowness of growth and structure." At germination, orchid

seeds require an external source of sugar because of the absence of energy stores in the seed. Once photosynthesis begins, the orchid plant requires nitrogen, phosphorous, boron, molybdenum and manganese. Under optimum light and heat, orchids will grow better — and often to a tremendous size not usually seen in nature — if they are constantly supplied with organic fertiliser. Experience in commercial nurseries has shown that even the inhibitory

effects of viral infection on plant growth and flowering can be repressed for several years by an aggressive fertiliser programme.

An early symbiotic association with mycorrhiza provides for the early nutrition of the newly germinating orchid seedling which is still unable to photosynthesise. From among the sugars (pentoses and hexoses) which are available in nature, the orchid seedling is capable of using only fructose, glucose and

Above: Intermediate *Dendrobium* (Hou Hin x Bertha Gold) in Bangkok. The plant is grown under saran which provides about 20 percent shade. This low-angled photograph features the saran screen as the background for the flowers. Note that the grower does not use a lath house for his *Dendrobium*. The plants are suspended from horizontal wooden beams that are about 30 cm apart.

sucrose. Once it has developed chloroplasts, the orchid produces its own sugar, but whenever a plant becomes shocked (for example, after prolonged dehydration), it will recover more rapidly if fed a dilute glucose solution.

Nitrogen is necessary for the formation of proteins, and a good supply of nitrogen is the key to rapid vegetative growth. The evidence suggests that initially the orchid seedling, notably in its protocorm stage, accepts only nitrogen presented to it as ammonium ions (HH4+) because it lacks the enzyme that reduces nitrates to nitrogen. This enzyme nitrate-reductase, appears as the seedling matures. Dr. Carl Withner found that orchids grow better when both ammonium salts and nitrates are supplied than if nitrates alone are provided. Nitrogen-fixing bacteria are also present on the leaves of all tropical vegetation as well as in the organic substrate on which the orchids grow, and these bacteria probably make some nitrogen available to the plant. Too much nitrogen, however, turns the leaves a dark green and prevents flowering. Phosphorus and potassium are more important for flowering and are also necessary for the development of tough supporting tissues.

The amount of fertiliser required by the plant and the balance of the nutrients are dependent on the species, the stage of growth, potting medium and growing conditions. When orchid seedlings are removed from the germination flasks, they have to be washed clean of the agar and then soaked in fungicide to protect them against soft rot. A dilute inorganic fertiliser and a small amount of glucose added to this fungicidal soaking solution will provide an additional boost. (Note that ants are not attracted to glucose.) Similarly, if seedlings are to be potted in charcoal chips, it is recommended that the charcoal be pre-soaked in dilute fertiliser overnight and thereafter the seedlings be watered once or twice a day and quarter- to half-strength fertiliser (NPK ratios of 4:1:1 to 2:1:1) be provided at least twice a week. The application of organic fertiliser (such as 'Nitrosol' or Fish Emulsion) once a month is also recommended. If provided with such fertiliser constantly in 50 percent sunlight and grown at 22 degrees Centigrade, seedlings of *Dendrobium* and *Phalaenopsis* can be brought to flower within 12–18 months.

After a year's growth, the seedlings may be toughened by gradually increasing their exposure to sunlight. The fertiliser should now be changed to one which contains a higher phosphorus and potassium ratio in order to encourage flowering. Continued application of high nitrogen fertilisers produces soft growth and delays flowering. In the monsoon regions, one should avoid high nitrogen, in particular organic fertilisers, during the rainy season and shift to high phosphorus and potassium if one wishes to fertilise the orchid at this time.

Once the plants are transferred to single pots, it is possible to add a fine sprinkling of bone meal, manure or pellets of controlled-release fertiliser to the pots. The liquid fertilisers can then be reduced. It is convenient to add liquid fertilisers to the spray only when insecticides or fungicides are being applied. Add bone meal every six months.

Perhaps the two most important factors which determine the need for additional nutrient supply to orchids are (1) the rate of growth of the plant and (2) the ability of the climatic or the cultural conditions to support a rapid pace of growth. In the cool, temperate countries, plant growth slows down in autumn and may come to a halt in winter. Watering should be reduced during this period, and feeding may even be discontinued. In the tropics, since high temperature and the abundance of sunlight support a luxuriant growth throughout the year, the vigour of the plant and its floral productivity are only limited by its supply of moisture and nutrients.

Opposite: A young *Phalaenopsis* plant established on the trunk of a *Podocarpus* (pine) tree in the author's garden. The orchid receives about an hour of direct mid-morning sunlight and dappled sunlight (about 50 percent shade) for the rest of the day. Within two years, the roots had spread all over the *Podocarpus* tree trunk, its tips a metre away from the stem.

Above: *Ascocenda* established on fibreglass 'rock' at the Singapore Botanic Gardens. They receive about 80 percent sunlight, including a few hours of full midday sun.

Right: *Aranda* Baby Teoh

Opposite: *Rhynchovanda* growing in the nursery of Nopporn Buranaraktham (Nopporn Orchids) outside Bangkok. This picture speaks a thousand words. With roots like this, one can be confident that the orchid plant is doing well.

In nurseries devoted to the production of cut flowers, the orchid plants are watered twice a day, liquid fertiliser being applied once a day or at least once in two days. In addition, every pot is liberally pasted with pig or poultry manure. When the *Aranda* is pumped with so much fertiliser, it flowers continuously around the year, but there is still a peak in summer. Vegetative growth proceeds apace with flowering. The plants can be top-cut and propagated once a year. The upper segments keep flowering even as they are re-potted, with only a modest reduction in floral output. Within six months, they attain the size of the mature plants, capable of maximum flower output. Because these plants are grown in full sunlight, without any overhead cover, leaching by rain necessitates frequent replenishing of fertiliser.

Potassium, phosphorus and magnesium are particularly important for *Dendrobium phalaenopsis* hybrids while *Cattleya* seedlings require a plentiful supply of nitrogen and phosphorus. Most *Aranda* and terete *Vanda* only give of their best when they receive organic fertilisers. *Phalaenopsis* are heavy feeders and are fond of bone meal; but as they are slow growers, only a tiny amount need be added once every two or three months. Species such as the *Spathoglottis affinis* go through a resting stage after they have flowered and should not be watered or fertilised during this period.

For a long time growers in the United States have noticed that orchids grown in osmunda do not seem to benefit from additional fertilising. Subsequent research showed that the osmunda fibre is almost identical in its mineral composition to the orchid plant although it is somewhat deficient in potassium. Nevertheless, in time, the nutrients in the osmunda fibre do get leached out, and at this stage, additional feeding with dilute fertilisers will give the plants a boost. Orchids which are grown on bark similarly require less fertiliser than orchids grown on brick and charcoal. Gravel and lightweight

clay aggregates do not provide adequate nutrients, and a balanced fertiliser programme is absolutely necessary for successful cultivation of orchids in a hydroponic system. It is the most difficult way to maintain orchids.

When the root system of the orchid plant is unhealthy, the grower must understand that the spraying of fertilisers on the plant and the addition of fertilisers to the potting medium do not result in any significant absorption of fertiliser by the plant. While the leaves may absorb some nutrients, they do so in amounts insufficient to maintain proper plant growth. To maintain healthy root growth, the pH of the water should be around 5.0. This is the run-off pH of bark, osmunda and charcoal. In areas of heavy rainfall, such as in the tropics, where orchids abound, ground water generally has a satisfactory pH, but if necessary, the pH of the water can be adjusted to 5.0 by the addition of phosphoric acid. Roots which have been drowned by excessive watering and root tips which have been burnt by strong concentrations of fertiliser or chemicals are also unable to absorb any nutrient. Many old plants have roots which are absolutely smothered by fungi and lichen which probably have first priority on any fertiliser running down the roots. Such plants will not thrive no matter how much fertiliser is sprayed on them, and they should be top-cut and the old roots trimmed off. For the sympodials, the first three lead bulbs should be removed and re-potted so that they can grow new, healthy roots. Old potting media must be discarded.

Chapter 7

Practical Aspects of Orchid Growing

In cultivation, an orchid passes through the following stages: the primary flask, the subculture flask, community pot, thumb pot, 8 cm pot, 15 cm pot, and finally the 30 cm pot, bark, or groundbeds. The requirements of seedlings up to the community pot size are fairly uniform for all species. Most growers will probably handle all their young seedlings in the same way, although each grower may vary the nature or amount of nutrients and stimulants which he adds to the agar, and he may select different media for his community pots.

It is only when the seedlings are placed into individual pots that their peculiar requirements call for variation in their culture. There is a wide range of recommendations for growing orchids from individual orchidists, each grower having his own recipes for achieving optimum results. Thus, instead of being placed in an 8 cm clay pot, the seedling may be grown in an 8 cm plastic pot, or in lightweight, open net pots, in teak baskets, or tied to a slab of tree fern, a piece of coconut husk or a stick of wood. There is also a large variety of potting media which may be used either exclusively or in combination (as mixes) — charcoal, broken clay bricks, redwood bark, tree fern chunks, osmunda, sphagnum moss, leaf mould and earth mixtures, Styrofoam, lightweight clay aggregates and gravel. Some growers prefer empty containers and do not use any media at all. The habit of the plant, and the presence

Opposite: *Vanda* Charles Goodfellow x *Ascocenda* Nopporn Gold. The plant was in superb condition when it produced this spray.

of ambient water are the two factors which determine the best cultural technique. In Bangkok, where orchids are grown by the banks of the Chao Praya River, it is customary to have water beneath the benches. One grower sandwiches his seedling between two strips of Styrofoam, while another throws them (without pot or medium) on the benches to await the arrival of new roots when they can be exported as thumb pot sized seedlings or planted into larger pots. The method is not used in Singapore, and certainly it would not work in a drier location.

In view of the diverse opinions regarding what essentially constitutes the best cultural conditions, and because many methods indeed give identically superior results, my method for growing *Phalaenopsis* will be described as a starting point. It is a simple method employing only two pot sizes and a single medium from community pot to adult plant. The other methods which are employed in this region will then be discussed. Finally, other cultural techniques more suited to the temperate countries will be reviewed briefly.

Seed Sowing

The best results for *Phalaenopsis* (and sometimes the only way to obtain hybrids from sub-fertile crossings) with regard to percentage germination are obtained through green pod culture. This is a technique of sowing the seed soon after fertilisation and long before the fruit ripens. Ripe seeds, especially with difficult crosses, have a poor germination rate and are often less vigorous. In addition, the longer pollination-sowing interval means a loss of time to the breeder, and there is always

the risk of accidental dehiscence of the pod.

The usual time for harvesting a green pod is 100–140 days after pollination, with the longer interval being preferred in cooler climates. The pod must be sowed within 48 hours of its removal from the plant because dehydration of the pod occasionally causes it to split open. When the pod is intact, the simplest way to sterilise its surface is to wipe it thoroughly with methylated spirit and (with some spirit still on the pod) to then put the pod to flame. The seed pod is cut into two or three segments with a sterile blade and sterile forceps, and the ends are trimmed off. Each segment is split longitudinally and the seed is scraped off into pre-sterilised flasks containing nutrient agar. About 10–20 ml sterile, deionised water is added before replacing the stopper. The simplest medium to use is Knudson's C formula, which can be purchased from laboratories and chemists, or directly from Difco Laboratories. To stimulate growth, 10 percent coconut water from young coconuts is added in place of the deionised or distilled water.

There are many methods to maintain a sterile environment during the sowing procedure, but the most convenient and reliable way is to make use of a laminar flow chamber. For the hobbyist who is only going to sow a dozen flasks per year, the cost of setting up a flow chamber may be prohibitive. Several nurseries advertise a sowing service in *Orchids*, the journal of the American Orchid Society, and it is best to make use of such services if one is only going to make an occasional hybrid. There are nurseries in many Asian countries that will also undertake sowing of orchid seeds.

Subculture

By the time the seedlings have produced their second leaf (usually six months after sowing), the primary flask would be crowded and the seedlings would have exhausted the nutrients in the agar. To allow for further growth, they should be subdivided and transferred into secondary flasks. An inexpensive subculture medium can be made up with 1.5 g of a regular inorganic fertiliser (Hyponex 20:20:20 or Gaviota 63) and 5 ml of Fish Emulsion, 20 g of liquidised banana pulp, 20 g of agar, 200 ml of young coconut water and 800 ml of water. Coconut water contains hormones which encourage rapid growth and callus formation which often result in bonus plants, while banana contains hormones which stimulate root formation. Tomato juice can be substituted if banana is not available, but its effect is in stimulating leaf formation rather than root formation.

A tremendous amount of material on culture media, orchid seed germination and mericloning is available in the series, *Orchid Biology: Review and Perspectives (I–VII)* edited by Joseph Arditti (earlier volumes published by Cornell University Press, Ithaca, USA; latest volume by Kluwer Academic Publishers, Dordrecht, the Netherlands). The interested reader should consult the series.

Seedlings in flasks are kept indoors or under 25 percent sunlight. After three months in the subculture flask, the plants can be moved out into the orchid house to acclimatise the seedlings to the light and heat which they will receive when they are transplanted from the flask.

Transplanting Seedlings from Flask

After they have settled in the flask the seedlings will show a spurt of growth, rapidly developing new leaves, each larger than the one before. After about six months, growth begins to slow down, and this is the correct time to remove the *Phalaenopsis* seedlings from the flask. By now, more than 80 percent of the seedlings should have developed at least two leaves which are 1 cm or more in length with two or three strong roots.

If you are planning to buy a flask from a nursery, choose one which has uniformly well developed seedlings rather than those which have a few large seedlings and clumps of protocorm.

Before unflasking the seedlings, prepare 500 ml of a dilute inorganic fertiliser solution (half-strength) for every 250 ml flask to be transplanted and collect together:

1. Three flat dishes
2. A single chopstick
3. A 25 cm length of wire of 1 mm thickness which should be bent at one end into a hook
4. A pair of pointed forceps
5. Six 8 cm plastic pots
6. Two grades of charcoal (medium grade pieces 3–5 cm in diameter and fine grade pieces that are 0.5–1.5 cm in diameter). I wash and pre-soak the charcoal in a dilute fertiliser one to two days before use.
7. Fungicide

To extract the seedlings from the flask, remove the stopper and pour in the dilute fertiliser so that the seedlings are submerged in the solution. After a few minutes, the agar will have softened, but if it does not, the chopstick can be used to gently break it up into several small parts. Shake the flask to dislodge the agar, and now the seedlings can be hooked out by the length of bent wire.

Occasionally, a flask may be overpopulated by large plants which are entangled in a clump. Rather than risk destroying some seedlings by forcibly pulling them out through the narrow neck of the flask, I prefer to break the flask with a gentle tap of a hammer.

The seedlings are washed in the dilute fertiliser solution to remove the agar clinging to the leaves and the roots. Three rinses are necessary to remove the entire agar. If the seedlings are clumped together, and especially if they still do not have any roots, they are best grown as a clump in the compot (community pot), and no attempt should be made to separate them at this stage. The seedlings are now sorted out according to size into three grades — large, medium and small. They are next transferred to a dish containing a fungicide, such as Captan, in addition to the dilute fertiliser. I rarely use a systemic fungicide, but should this be preferred, the seedlings should be left to soak for 30 minutes to allow the fungicide to be absorbed into the plant tissue.

The best pot for growing seedlings in communities is the 8 cm plastic pot, although commercial growers dealing with very large numbers of seedlings sometimes prefer 20 cm clay pots. All pots must be spotless. The bottom half of the pot is filled with three to four pieces of charcoal 1–2 cm diameter and thickness, and the large seedlings are planted in a circle around the periphery, using smaller pieces of charcoal (cubes 1–2 cm in diameter) to anchor the plants. A pair of pointed forceps is helpful in handling the charcoal. Several seedlings are now tucked into the centre of the pot and anchored with more charcoal. All the leaves and stems must be above the surface of the potting medium. When fully planted, the surface of the charcoal should be 1.5 cm below the rim of the pot. I find this helpful in maintaining humidity around the seedlings.

If the seedlings are large, three or four seedlings may be planted in a compot. Generally, 12–20 medium-sized seedlings can be placed together in an 8 cm pot. If the seedlings are tiny, very fine charcoal (about 3–5 cm in diameter) has to be used for planting, and up to 50 seedlings can be placed in the 8 cm pot, to be redivided into several community pots as the plants get larger.

A label is now prepared for each community. This should contain the essential information on the parentage of the seedlings. Many growers use a numerical code for speed and simplicity. The date of transplanting and each subsequent transfer to a larger pot can also be recorded on the label to give the grower some idea of how fast the plants are growing. The cheapest labels are made from thin plastic strips but a more durable label is commonly available from most nurseries. I use a white label for all seedlings and colour-coded labels for selected plants.

My community pots are grown crowded together, usually nine pots or more at any one time. If I happen to have only one or two community pots from a small flask, I like to hide them among larger established seedlings so that they are protected from dehydration by the high humidity of their surroundings. I have obtained the best results by placing the seedlings in a box lined with PVC and layered with a 5 cm thickness of lightweight clay aggregates beneath which is a centimetre of water. The seedlings must be sheltered from rain and can receive 40 percent sunlight initially. They are finely sprayed once a day and given fertiliser twice a week.

After 4–6 months, each seedling will have at least two new leaves and several strong firm roots. When the average plant is 6–8 cm across from leaf-tip to leaf-tip, it is time to separate them into individual pots. However, if the compot has become very crowded well before the seedlings have reached this size, I divide the crowded community pots and grow a smaller number of the seedlings in freshly reconstructed community pots. Similarly, when separating any community pot, there

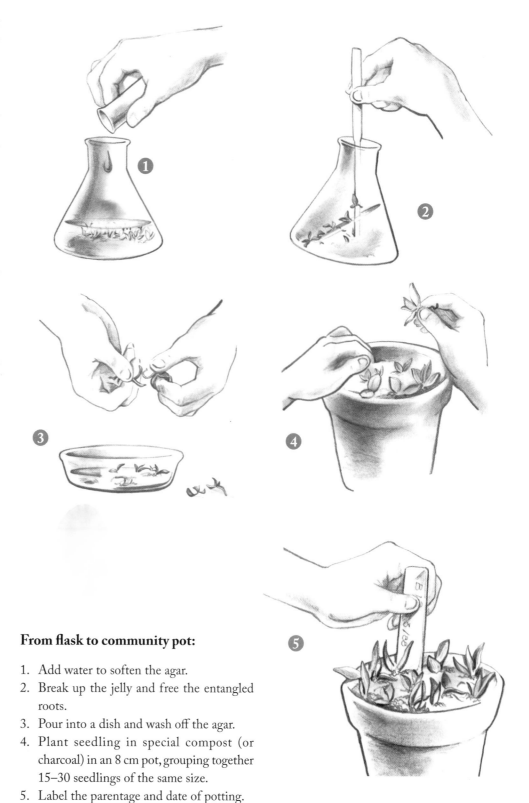

From flask to community pot:

1. Add water to soften the agar.
2. Break up the jelly and free the entangled roots.
3. Pour into a dish and wash off the agar.
4. Plant seedling in special compost (or charcoal) in an 8 cm pot, grouping together 15–30 seedlings of the same size.
5. Label the parentage and date of potting.

will always be a few smaller seedlings which do not appear ready to strike out on their own and they (three to five plants) should be grown together.

To remove the community of seedlings, the pot is dipped in water to wet the potting media and all the roots. After a few minutes, the plants can be dislodged by turning the pot sideways and tapping on the bottom. Though the plants come off easily from the pot, their roots may stick to pieces of charcoal. The larger plants should be individually and carefully separated. Smaller plants which are still stuck together can be repotted as a group. Most growers step down to thumb pots (small 3 cm pots), but I use the same 8 cm plastic pot for individual plants as they can remain in this pot until they flower. Charcoal is used as the sole repotting medium, with larger pieces being used to pot the larger single plants. The individual pots are returned to the same bench used to grow the community pots, and the plants are watered and fed with the same frequency. After eight weeks, these newly potted plants will be established, and a small amount of bone meal (such as Gaviota Orchid Fertilizer) is sprinkled over the charcoal to provide the plants with a constant source of fertiliser. Application of fertiliser is repeated once every six months. As the seedlings grow bigger, the pots need to be spread further apart, and the plants are now exposed to 50 percent sunlight. This is achieved by removing the strip of plastic netting on the lath roof above the seedling bench. My seedlings are allowed to grow in the 8 cm pot until they have produced at least two new leaves of a minimum length of 10 cm each. They are now ready for the final transfer to the optimum size, 15 cm plastic pots. By now, they have been out of the culture flask for 12–18 months.

Larger pieces of charcoal need to be used for plants in 15 cm pots to prevent the roots from becoming waterlogged. I select a fairly large piece of charcoal, 12–15 cm long, to place

diagonally across the pot with the plant placed astride this piece of charcoal. Two or three slightly smaller pieces of charcoal are placed across the first piece to fill up two-thirds of the pot, and finally, 3 cm cubes of charcoal are placed on top. Care must be taken to ensure that the stem of the *Phalaenopsis* plant is always above the surface of the charcoal. When they are returned to the bench, the plants should be placed as close together as possible but avoid any overlap of the leaves. When the plants are re-established after 8–10 weeks, the pots can be rearranged to provide more growing space for the individual plant and more air movement around the plant.

As a rule, the first plants of a hybrid to flower do so 18 months after unflasking.

The Cultivation of *Dendrobium* and Other Sympodial Orchids

Dendrobium are sympodial orchids. Most members grow upright while a few members (belonging to the section *Eugenanthe*, such as *Den. nobile*, *Den. pierardii* and *Den. anosmum*) produce pendulous stems which hang downwards. Both types can be grown attached to living trees, tree stumps, rafters and tree fern slabs. The common way to grow erect *Dendrobium* is in clay pots or similar containers.

Growing in Containers

The best containers for *Dendrobium* are perforated clay pots, which will provide excellent aeration for the roots. They will grow more rapidly in unperforated clay pots or in plastic pots because of the higher humidity around the roots, but the plants become pot-bound very much sooner and will thus require more frequent repotting. Wooden baskets are also less satisfactory because they are difficult to handle and the plants require more watering.

Pot size depends on the size of the plant at the time of potting, with provision for the size that it is likely to attain in one to two years. The usual pot sizes range from 8 cm to 30 cm, 15 cm being the commonest size for mature plants. Never use an excessively large pot for a plant as this will not encourage it to grow faster. In fact, the plant may not thrive when it is overpotted, and the practice is wasteful of space, potting material and chemicals. Thai growers like to keep their plants in small pots. Start with the smallest pot that can hold the plant. As it outgrows the pot, just place the plant in a pot of the next size, and so on. In Bangkok, one can see robust plants growing in tiny 8 cm pots, their roots holding tightly to a flattened layer of tree fern and gliding over the rim of the pot to attach themselves onto the outer surface.

Most *Dendrobium* are potted with 3–4 bulbs, and at the end of two years, if the plant has been well grown and properly divided, there should be a dozen bulbs, and it is ready to be separated into four separate plants. If one is planning to grow the *Dendrobium* into a large specimen plant, one still has to start with a 15 cm pot, gradually changing to a 30 cm and then 40 cm pot in two transfers, about 12–18 months apart. In this way, the soured old potting medium is discarded with each repotting, and the roots remain healthy to sustain the specimen plant.

In Singapore, the commonest potting medium for *Dendrobium* is broken charcoal, used exclusively or in combination with broken brick in various proportions (the ratio is not critical but is usually 4:1, charcoal to brick). If the *Dendrobium* are to be grown as hanging plants, using charcoal exclusively will lighten the load on the lath beams. On the other hand, if they are to sit on benches, the heavy brick at the bottom

of the pot gives it stability. The bottom layer (30 percent of the depth) is made up with 4–5 cm cubes of charcoal or broken brick. For the next layer, 2–3 cm cubes of charcoal are used, and finally, the empty spaces at the surface are covered with the smallest pieces of charcoal, each approximately 1 cubic cm. One could use redwood bark for the uppermost layer to provide additional dampness and acidity as this encourages faster root growth, but this is expensive and, on a large scale, not a justifiable expenditure.

Opposite: Community pots (compots) of *Phalaenopsis* seedlings

Above: This *Dendrobium* has outgrown its pot and with the three new shoots appearing, it is time to repot. The old canes look unhealthy and should be separated and left unpotted to see whether they will produce offshoots.

In Thailand, the pots are filled three-quarters to the top with crock, and the last quarter is packed with vertical fibres of tree fern which are trimmed flush with the top of the pot. The *Dendrobium* canes are rested on top of the tree fern, and their roots grow in between the fibres.

The correct time to repot *Dendrobium* is when a new pseudobulb is just beginning to put forth its own roots. Two or three mature pseudobulbs should be present behind the young pseudobulb to provide it with adequate nutritional support after repotting. It does not matter if these back bulbs are actually smaller than the new bulb, for such would be the case if the plant were a young seedling. If the plant is extremely large, it can be propagated by removing only the young pseudobulbs that are supported by three back bulbs and leaving the rest of the plant in the old pot. When the growing lead is removed, the inhibitory effect which it had exerted on the axillary buds (the 'eyes') in the old back bulbs is eliminated, and within a few weeks, new shoots will develop from these resting buds. If space is critical, the back bulbs may be removed from the pot and suspended in a cool spot on the side of the lath house after trimming off its roots. When a new shoot appears, it will develop its own roots, and the back bulb, with its new shoot, can then be repotted.

It is customary when potting *Dendrobium* to position the plant halfway between the centre and the rim of the pot, with the growing tip facing inwards so that there is ample room for the plant to grow forwards and to branch sideways. The plant must be firmly anchored, and this is done by tying the canes either to stiff wires which are used to suspend the pot from the lath roof or to bamboo sticks staked firmly into the pot. In no event should the pseudobulbs be buried in the potting medium. The pseudobulbs must sit just at the top of the potting medium with the roots anchored in the medium.

Newly repotted plants require 50 percent shade and fine spraying twice a day until the roots have developed and anchored the plants securely to the pot. They are then gradually exposed to between 80 percent and full sunlight. All *Dendrobium* do well under 80 percent light in the tropical lowlands, but cane *Dendrobium* (members of the *Ceratobium* group) and their 'intermediate' progeny flower better with full sunlight. For growing under shade, lath houses are most satisfactory from the point of view of maintenance and appearance, as the plants are usually grown in hanging pots. An alternative method of housing, which is particularly suited for growing *Dendrobium* in the open, is to place them on skeletal benches made up of two parallel, horizontal wooden beams, supported 60 cm above the ground. The distance between the parallel beams is adjusted so that they will grip the pot at its waist, just below the raised rim. The pots must be of uniform size and manufacture to make proper use of this system. These benches are constructed to hold three rows of pots such that the centre row is still within easy reach of the grower.

Growing on Tree Stumps and Slabs

Because their habit is to hang downwards, the members of the section *Eugenanthe* and their hybrids (known as the *nobile* type) are more easily grown by simply tying the canes on to a stump of wood or a slab of tree fern and covering the roots with osmunda. A convenient material for tying the plant without risk of injury to the roots is plastic twine or rubber-coated electric wire. A small nail should be driven into the wood to anchor the twine. If osmunda is not available coconut husk may be substituted. Yet another method employs a small sheet of plastic netting wrapped around some bits of tree fern, which is used to surround the roots. The newly planted *Dendrobium* must be stood in a shady spot and watered at least twice (or preferably thrice) a day until it is established, after which the shading can be reduced.

If it is the intention to grow *Dendrobium* on living trees or on huge stumps of wood, they should first be established on slabs of tree fern or on short stumps of wood. When the plant has developed at least three large pseudobulbs, they are collected together and tied or nailed to the trees as part of the landscape. The best time to do this is at the start of the rainy season. Remember that their requirement is 80–100 percent sunlight, so they should not be grown under short, densely leafed trees. The favourite tree stump in Singapore is the main stem of a fallen mangosteen tree because of the durability of its wood.

Cultivation of Monopodial Orchids — *Vanda-Arachnis* Tribe

For purposes of cultivation, all monopodial orchids can be divided into two groups: (1) ground orchids whose roots require the high humidity and humus content of the earth and (2) tree-borne epiphytes which are derived from those species which grow perched on the branches of tall trees. Monopodial ground orchids are generally all sun-loving. The tree-borne epiphytes may be subdivided into sun-loving and shade-loving groups, and their cultural requirements are different. The sun-loving epiphytes require good ventilation around their roots and can withstand dry spells quite well, while the shade-loving epiphytes, like the *Phalaenopsis*, can be grown within an enclosed pot, but they require a high humidity at all times.

The Cultivation of Ground Orchids

Ground orchids can be grown in beds in the earth or in pots. There are two main reasons for growing orchids in beds: (1) for landscaping, and (2) to grow large numbers of a type for the flower market. In order to create the visual impact in landscaping, the orchids in a single bed should preferably be of one type and of one colour. If one has only one or two plants of a type, it is best to continue growing them in pots.

Almost all Spider Orchids (*Arachnis* and their intergeneric hybrids), *Renanthera* and their intergeneric hybrids, terete and semi-terete *Vanda* do extremely well if they are grown on the ground in full sun with liberal watering and fertilising. Sloping or flat ground with good drainage provides the ideal location for orchid beds. The land should be cleared of grass and all ground cover to make a bed 2 m by 0.8–1 m. If the space is insufficient for this bed size, it is better to grow the plants in large clay pots. Vertical stakes of wood, 4.5 cm by 4.5 cm to 6 cm by 6 cm in cross-section, are driven into the ground in pairs, each 0.5 m from the other, at 2 m intervals along the length of the bed. The stakes should penetrate the ground for a depth of 30 cm and be 1.5–2 m above the ground. Horizontal wooden beams 3 cm by 3 cm in cross-section are now spliced to the stakes at 30 cm intervals above ground level. For a start, only two horizontal beams need to be positioned. Finally, wood shavings, to a depth of 6 cm, are deposited over the bed as ground cover. The bed is now ready for planting.

The best size for planting out in beds is when the orchids have reached a height of 70 cm, as they are then robust plants, ready to flower. A seedling which has been originally grown in a pot can be completely transferred to the ground, but the common practice is to take a top cutting, leaving behind a bottom stump of 10 cm with several leaves which will then become the source of bonus plants. Make sure that each cutting has at least two strong roots. Some orchids, such as *Renanthera*, can grow very tall and remain dependent on their bottom feeding roots for water and nourishment. The stem may be completely devoid of roots for perhaps a metre from the growing tip. Such plants may have to be completely unpotted for planting in the ground. Try to start off a bed with cuttings of uniform height.

The bottom end of the stem should preferably be sealed with tar. Some growers dip it in Tersan or a similar fungicide and allow the end to seal up through dehydration. Even if nothing is done to the cut end, it will usually seal up, and it is only in rare instances that fungus will attack the ground orchid through the cut end of the stem.

Stand the cuttings along the horizontal wooden supports, allowing a 5 cm distance between leaf tips, and anchor them to the wood with plastic twine which will not cut into the stem. The distinguished Malaysian grower, Datuk Dr. Yeoh Bok Choon, suggests that the bottom tip of the stem should be 15 cm above ground level to separate it from the damp, spore-ridden ground cover. The roots are now tucked into the wood shaving and a further 10 cm depth of wood shavings, is heaped above the roots. Cut grass was once popular as alternative ground cover for the roots but it may inadvertently introduce some pests to the orchids.

The newly planted orchids require some shade from direct sunlight. Even as little as 15–25 percent shade will usually suffice, and this can be provided by a temporary plastic or saran netting. It must not be removed for at least eight weeks or until the plants are established on the pots. The newly planted orchids should be watered twice a day, morning and evening, and fertilised with dilute fertiliser (preferably organic) at least twice a week. It is a good idea to put out the orchids into the beds at the start of the rainy season in order to take advantage of the rain, if cloudy, rainy days are guaranteed for several weeks and overhead cover may be dispensed with.

When the plants are established, controlled-release fertiliser or powdered bone meal can be scattered over the ground cover. Ground orchids are extremely hardy and thrive on fertiliser.

If the stakes are 2 m above the ground, they will provide adequate support for another two to three years' growth by which time the plants ought to be pruned and propagated. Top cuttings take less time to bloom than offshoots, and the two should be planted separately.

Until the mid-1950s, *Vanda* Miss Joaquim was still popular as a garden plant in Singapore and Peninsular Malaysia, and the method for bedding was to tie cuttings around vertical wooden stakes 8 square cm in the cross-section. The bottom halves of the plants which bore the feeding roots were covered by a thick paste of cattle manure, and the usual ground cover was provided by grass cuttings. All terete *Vanda* will respond well to this treatment, and it can still be recommended for those areas which have easy access to cattle manure.

Cultivation in Pots

Monopodial orchids can be grown in pots; in fact, this is the safest method for growing a precious plant. Tall-growing monopodial orchids are best grown in large clay pots, up to 30 cm in diameter, which provide proper anchorage for the plant as it clambers upwards.

There is no minimum size for the plant to attain before it can be repotted into a clay pot, but the rule of two strong healthy roots applies. Naturally, the more roots to the stem and the taller the stem, the less time it will take for the plant to re-establish itself after repotting. However, as a rule, if repotting involves top-cutting, primary consideration is given to the backshoot, which should be left with at least six leaves to allow for an optimum yield of offshoots. Although the top-cutting will also, in turn, be capable of producing offshoots, one has to wait a longer time first for it to establish its own roots and then for the plant to be ready for top cutting.

If the roots are long, first soak them in water for a few minutes to render them pliable and then place the plant into the new pot. The roots can now be curled into the pot. A slim pole, 1 cm in cross-section, is tied to any plant which is likely to grow beyond a metre in height, and plant and stick are tied together with plastic twine and firmly anchored by small pieces of broken brick. The broken brick normally fills one-third to half of the pot. Broken charcoal is now used to fill the pot up to within 2 cm of its rim. Make certain that the supporting stick and the plant are steady and upright.

Styrofoam, tree fern chunks and redwood bark have been used as alternative potting media but have not gained wide acceptance. Styrofoam does not retain any water, while tree fern and redwood bark tend to cause waterlogging.

The potted plant is grown in a shady spot which may receive 50–75 percent sunlight, depending on the size and vigour of the transplanted plant, until its roots are established and the plant is then gradually moved into full sunlight. It is customary to place the pots in 2–3 rows on raised benches which are 60–80 cm above the ground. Watering and fertilising should follow the pattern for bedded monopodial orchids.

If the orchid has short leaves, it is economical to tie 2–4 plants to a single post and grow them together in a single pot. This enhances the impact when the plants flower simultaneously.

Shade-loving monopodial orchids, such as the *Phalaenopsis* and their intergeneric hybrids,

Growing in Baskets

The simplest way to grow strap-leaf *Vanda*, *Ascocenda*, *Rhynchostylis* and *Aerides* is in hanging baskets, although they can also be grown in pots, on trees and even in the ground. Square teak baskets are available in various sizes with side lengths varying from 8–20 cm, the universal size being 15 cm. These are suspended in rows, 20 cm below eye level, in lath houses. Orchids are gregarious and only happy when they have the humidity generated by a good number of plants grown in close proximity. The rows of hanging baskets are commonly set 30 cm apart and 1.2 m deep to allow easy access on either side.

A robust thumb pot seedling can be transferred to an 8 cm basket simply by placing the pot into the basket, leaving the roots to anchor the plant to the basket. When the plant has overgrown the 8 cm basket, simply remove the supporting wires and drop the plant into a basket which is one or two sizes bigger. This process can be repeated once more when the plant has grown to a very large size.

A bare root cutting needs a little more support when it is newly placed into a basket. The simplest way to anchor the plant is to tie the base of the stem to the central beam at the bottom of the basket and to pass some of the roots through the spaces between the horizontal beams. Further up, the stem is anchored to the four supporting wires of the basket by elastic twine. No potting medium is required, but if one likes to have stronger support at the base, a few large pieces of charcoal can be placed in the basket to provide additional anchorage for the roots.

On these two pages: Some imagination is required to produce an attractive display when orchids are planted in the ground in one's garden. These two examples from the Mandai Gardens in Singapore show how good display can be achieved with curves and circles. Its original owners, John Laycock, Lee Kim Hong and Rosalind Lee and John and Amy Ede were trailblazers for landscapers in Singapore.

have to be grown in 50 percent shade, and as mentioned earlier, they are best grown in plastic pots in straight charcoal.

Chapter 8

Propagation

Carbon copies of an orchid can only be produced through vegetative propagation, which can be achieved by four methods:

1. Division of the plant, otherwise known as cutting
2. Shoot development on old back bulbs
3. Stimulation of plantlet formation on flower stems
4. Meristem tissue culture

Some *Phalaenopsis* species may, on rare occasions, produce plantlets at their root tips.

For the average gardener, simple division and raising plants from back bulbs provide the two simplest methods to increase the collection. *Phalaenopsis* produces only 3–4 leaves per year and cannot be quickly propagated by cuttings, but a hormone paste has been developed which readily stimulates production of plantlets on old flower stems.

Cuttings

This method is suitable for both monopodial and sympodial orchids. In the case of monopodial orchids, the optimum time for cutting is (1) when the top-cut portion can have at least two strong roots, and (2) the bottom segment is left with at least six leaves and four unflowered leaf axils. Make a clean cut with pruning shears, not with a knife. Seal off the cut ends with coal tar or a fungicide, and allow the tip to dry before replanting the top-cut. After a few weeks, the bottom segment will produce several new shoots and when these have developed two or three strong aerial roots, they may be removed and replanted.

If the aim is to propagate the plant, it should be given 50 percent or even heavier shade, liberal watering and heavy nitrogen feeding. This will stimulate rapid soft vegetative growth and prevent flowering. Take care to spray fungicides at close intervals, because under such growing conditions, the plant is prone to bacterial and fungal rot. When making a top-cutting, more of the stem is left behind in order to obtain additional offshoots. If the offshoot is very strong and there are no more axillary buds left on the main stem, the offshoot need not be removed completely, but instead a short stump of the offshoot bearing two or four leaves can be left behind on the old stump; this in turn will yield another generation of offshoots.

A higher yield can be obtained by making multiple cuts on the stem, but this can only be done if the plant has been grown so that its roots are firmly attached to a broad stump of wood. Each division should have six leaves, and they are all left attached to the original support. Shading is increased, and liberal watering and heavy nitrogen fertiliser are given. After a few weeks, three or four offshoots will develop from each segment. When these are strong enough, they may be removed and repotted.

If the orchid shows no attempt to produce aerial roots, it can be stimulated to do so by making a cut of about one-third of the diameter of the stem. A tall, lanky plant can be forced to bend, and below the point of flexion, it will usually produce several offshoots.

In the case of sympodial orchids, the following condition, should be met before dividing the plant:

1. It should be healthy. (If the plant is weak, it is better to remove a back bulb.)
2. It should have at least four mature stems.
3. A few pseudobulb, should be developing from the base of the old one and just beginning to send out its own roots.

A clean cut is made in the rhizome, leaving at least two mature pseudobulbs attached to the young pseudobulb to provide it with nourishment. After sealing off the cut ends with tar or fungicide, the divided plant can be left in the pot, or the younger segment can be removed and repotted. The back portion will produce at least one new pseudobulb if it is healthy.

If the original plant is large and extensively branched, several cuts can be made. Each segment should have 2–3 strong pseudobulbs. After a few weeks, several young pseudobulbs will develop, one or two behind each cut on the rhizome. If the plant is left in the pot, it will produce a spectacular flowering when all the new pseudobulbs send off their blooms. When the pot is overcrowded, the entire clump should be removed, and each division identified, freed and repotted. Most growers trim off all the old roots, but some will leave the old roots behind if they are healthy, since they are still capable of absorbing water and nutrients. Clear off all the old potting medium and soak the roots in fungicide before repotting.

Treatment of Back Bulbs

Instead of removing the new lead as described above, one or two old pseudobulbs at the end of the rhizome may be removed. This may actually strengthen the plant left in the pot. The pseudobulb is hung up under shade and watered occasionally until a new shoot develops, when it can be replanted in the usual way. If the bottom-most bud is healthy, it will be the one to develop into a new shoot; otherwise any bud along the length of the pseudobulb may develop into a new plant. In damp tropical regions, there is no advantage in tying *Dendrobium* back bulbs to a piece of wood or tree fern. However, in cooler, drier climates, various methods are employed to increase the humidity around the back bulbs

to encourage the development of offshoots, such as layering on peat and sand or dropping them into a polythene bag which contains a small amount of moist sphagnum moss or osmunda.

Dr. Joseph Arditti of Irvine, California, has developed an aseptic technique of planting segments of *Dendrobium* pseudobulbs, where each contains an eye, in enriched nutrient agar. A single pseudobulb will yield perhaps a dozen plantlets or more and, presumably, as new uncontaminated pseudobulbs grow from the eye, they could be redivided and propagated in the same medium. (Refer to the *American Orchid Society Bulletin*, 1974, Vol. 43, p.1055 for details.)

Opposite: *Dtps.* Luchia Pink. This beautiful orchid is readily available and inexpensive because it is an Asian mericlone.

Below: The old canes on *Phalaenopsis*, cane and intermediate *Dendrobium* frequently produce offshoots that can be removed and replanted when they have sufficient roots.

Plantlets from Flower Stems

As far as I am aware, only the flower stems of *Phalaenopsis* and its intergeneric hybrids and the old inflorescences of equitant *Oncidum* produce plantlets spontaneously; although, very rarely, the flower stems of *Dendrobium* may become vegetative and turn into a plantlet. This habit is commonly exhibited by the pink and star-shaped *Phalaenopsis* of the Philippines, such as *Phal. schilleriana* and *Phal. lueddemanniana*, and by the widely distributed *Phal. cornu-cervi*. Old flower stems of such species should always be left on the plant after flowering, for they are frequently the source of bonus plants. The same propensity is exhibited by their hybrids.

In 1962, Dr. Yoneo Sagawa of Hawaii found that when the internodes of *Phalaenopsis* flower spikes were sterilised, a good proportion of them developed into plantlets while some produced flower spikes. The method works particularly well with plants of *Phal. lueddemanniana* bloodline but is more difficult with white *Phalaenopsis*. Then, in 1971, a Taiwanese orchidist developed a paste which caused *keikis* to develop on the flower spike itself. The method was not published, and I could not obtain the formula from the people who were using the paste. However, I found that when I made an aqueous cream with 0.05 to 0.5 percent benzyladenine (6-benzylamino-purine) and applied it to old flower spikes of *Phalaenopsis* and *Doritaenopsis*, the cream caused the axillary buds to swell and develop into single plants or clumps of plants. This works equally well with pure white *Phalaenopsis*.

Benzyladenine can be made into a paste for application on flower spikes by mixing it with either aqueous cream or lanolin. Have your

Left: Chemically(BAP)-induced *anaks* on *Phalaenopsis* old flower stalks

chemist do this for you. To obtain a 0.05 percent concentration, mix 0.5 mg of benzyladenine with 1 g of cream or lanolin. Select only old flower spikes which have finished flowering a few weeks earlier. After removing the bracts from the two internodes beneath the first flower, apply the paste lightly to the bud. These axillary buds will swell after four days, and the first leaves will break through after a fortnight. After 2–3 months, some plantlets will develop roots; a few may take a longer period. The leaves of these plantlets are very succulent, and they require a lot of water, so the plants must not be removed for repotting until they have developed strong roots. It is a temptation to apply the paste on all the nodes on the flower spike, but this will probably weaken the parent, plant and the benzyladenine may possibly inhibit the growing point on the plant. It is safer to produce only one or two plantlets on a single inflorescence. On the other hand, if the parent plant is dying from crown rot or weevil attack, then one may feel free to treat all the axillary buds.

Meristem Tissue Culture

Meristem tissue culture is a major milestone in orchidology, because through this relatively simple method, anyone can achieve a rapid propagation of selected orchid plants. It has already been widely applied to produce selected clones of *Cymbidium*, *Cattleya*, *Dendrobium*, semi-terete *Vanda* and *Aranda* for the flower trade. Hybridisers have also taken advantage of the process to multiply valuable parent plants, particularly the polyploid forms of *Cymbidium*, *Cattleya* and *Phalaenopsis*. The other benefits of the method include the production of virus-free plants, new polyploid forms and peloric forms.

The term meristem refers to the actively growing, undifferentiated tissue which is found at the growing tips of plants. Meristems can be obtained from shoot tips, young inflorescences, axillary buds and root tips. Young, developing leaves also possess meristematic tissue. Dr. George Morel, working in Paris, found that when meristematic tissue was removed from the plants and placed in liquid nutrient media, it grew and multiplied but remained undifferentiated if the effect of gravity was eliminated by continuous agitation of the culture. He used this method to remove uncontaminated meristems of potato and Dahlia to obtain virus-free clones. Subsequently, Morel was persuaded by Michel Vacherot to work on *Cymbidium*. Vacherot Lecoufle et Fils became the first commercial orchid nursery to offer meristems of awarded orchid plants to orchid growers. Meristem culture of orchids has been further developed and reported in orchid magazines by Donald Wimber, Yoneo Sagawa and Robert Scully Jr. Improvements of culture media have been extensively investigated by Joseph Arditti.

Essentially, the method of meristem tissue culture involves the removal of a 2 mm cube of meristematic tissue which is inoculated into a liquid culture medium under sterile conditions. This is sealed in test-tubes which are placed on a rotary shaker or into small conical flasks which are agitated on a horizontal shaker. According to published reports, active meristems from shoot tips will continue growing and developing into large protocorms in the simple Knudson's C medium from which agar has been omitted but to which 20 percent of young coconut water has been added. Cytokinin, NAA (naphthalene acetic acid) or IBA (indole-butyric acid) is frequently added to accelerate growth of the tissue. Dormant meristems will benefit by the addition of benzyladenine. When the protocorm has attained a considerable size, it is divided and transferred to fresh medium, and the process is repeated until a sufficient number of pieces have been obtained. Thereafter, they are planted into solid agar medium, where the protocorms will develop leaves and roots.

The use of growth stimulants on the meristem appears to accelerate their growth beyond that of ordinary seedlings. Meristemmed *Dendrobium* can flower in less than a year after planting out from flasks, and *Aranda* in two years. Meristemmed plants are usually true copies of the clone, but a significant percentage show variation in size and substance of the flowers, width of the floral parts and arrangement on the rachis when grown under identical conditions. Polyploid forms may be found among the meristemmed plants, probably the result of endo-reduplication, together with aneuploid forms and possibly some chimeras. These two latter varieties are recognisable by their abnormal flowers.

When purchasing a meristemmed plant, try to buy one that is in bloom, and pick the best flower within the group. Do not go by name alone.

The technique of meristem tissue culture has been widely described and can be easily developed by anyone who has the facilities for seed sowing. Several firms supply various culture media in convenient packing; rotators are inexpensive and the use of a laminar-flow chamber practically guarantees sterility. However, it is a time-consuming process; protocorms need periodic transfers to fresh media. Most growers will find it more convenient to send their plants to laboratories which are now providing a meristem service. In large quantities, the fees are reasonable. If you are living in the tropics, your plant should be meristemmed in a laboratory located in the tropics (Thailand, Singapore or Hawaii) because the plant should be growing vigorously when the meristem is taken. These laboratories are probably also more familiar with the requirements for the genera which you grow.

Chapter 9
Hybrids

In the island republic of Singapore and in Peninsular Malaysia, commercial orchid ranges with acres of orchids under cultivation are devoted entirely to orchid hybrids. With the strict conservation laws in practice in many countries, it appears that pure orchid species will gradually disappear from amateur collections. The overwhelming popularity of hybrids is due to a variety of reasons: ease of cultivation, their free blooming habit, compactness and the fantastic array of shapes and colours. For the orchid enthusiast, there are over 75,000 different hybrids to choose from, the result of just 120 years of horticultural effort.

It all started in 1856 when John Dominy flowered *Calanthe* Dominii, which he had created by crossing *Calanthe masuca* with *Calanthe furcata*. A simple gardener with no botanical training, John Dominy did not appreciate the theoretical barriers to interspecific breeding. Instead, he acted on the advice of a surgeon, Dr. John Harris, who showed him how to transfer pollen from one orchid to the stigma of another. In recognition of Dr. Harris' contribution to orchid hybridisation, Professor Reichenbach linked his name to the first *Paphiopedilum* hybrid, *Paph.* Harrisianum, a cross between *Phal. barbatum* and *Phal. villosum*. It was a formidable task to raise seedlings from seed at that time, and when Dominy retired in 1880, he had created only 25 hybrids. Many hybridisers of that period who reported making crosses never managed to raise them to blooming size, and it was only in 1873 that another successful breeder appeared. He was John Seden, who took over from Dominy as chief gardener at the Veitch Nursery. Between 1873 and 1905,

Seden made over 500 hybrids, including the *Phalaenopsis* John Seden, a cross between *Phal. amabilis* and *Phal. lueddemanniana*.

Almost from the start, a careful record was made of all the orchid hybrids. The first list, compiled by F.W. Burbidge, appeared in the *Gardener's Chronicle* in 1871. Modern hybridisers refer to the Sander's list of orchid hybrids, together with its supplements and addenda, which provide a complete, systematic listing of all the hybrids right down to the present day. The first *Sander's Complete List of Orchid Hybrids* appeared in 1946, and the three-yearly addenda which brought the list up to 1954 were the life work of Frederick K. Sander of Sander's, St. Albans, England, who devoted more than 50 years to their compilation. The work was continued by the Registrar of Orchid Hybrids of the Royal Horticultural Society in London. Each month, a list of new hybrids appeared in the *Orchid Review*, and this was reproduced across the Atlantic in the *American Orchid Society Bulletin*. Every five years, a new addendum was added to the Sander's list. The RHS replaced the book form of the Sander's list with a CD 10 years ago — a single CD holding the data from 6 volumes of the list. Now, it is also possible to check the parentage of an orchid on the website.

If you have made and flowered a new hybrid and wish to give it a name, it must be properly registered with the Registrar of Orchid Hybrids, Royal Horticultural Society, Vincent Square, London, S.W.1. A description of the flower on a prescribed form or a coloured slide should be sent to the registrar together with the prescribed fee. (Determine the amount with the Royal Horticultural Society.)

Registration of a New Hybrid

In order to qualify as the Breeder of the Plant you must be the owner of the pod (seed-bearing) parent when the cross was made. It does not matter if you did not make the pollination yourself: ownership of the pod-bearing parent determines the breeder status. If you had used a plant belonging to a friend to make the hybrid, you should enter into a lease agreement whereby the friend grants you in writing the right of temporary ownership or co-ownership of the pod-bearing parent. Such leases should expire when the seed pod has been harvested. A breeder should keep a Stud Book, which should contain the date of pollination; the parentage of the cross; the date of sowing, the result (whether germination took place and whether plantlets were produced); dates of subculture(s); planting out, first flowering, the name of the hybrid and, finally, some comments on the results.

Opposite: *Ascda.* Kwa Geok Choo
Below: *Ascda.* Kwa Geok Choo x *V. denisoniana*

Classification of Hybrids

There are three types of orchid hybrids:

1. A *Hybrid Species* is produced by selfing a species or by crossing two plants of the same species.
2. An *Interspecific Hybrid* is produced by crossing two different species of the same genus, or by secondary crossings with other interspecific hybrids.
3. An *Intergeneric Hybrid* results from crossing of orchids belonging to different genera.

When hybrid names first appeared, they were identified by a cross in front of the name, such as *Calanthe* x Dominii. Today, we write hybrid names in plain lettering, with capitals for each word in name, for example, *Vanda* Tan Chay Yan or *Phalaenopsis* Penang. Species names are written in italics and do not begin with capitals, as in *Vanda hookeriana* or *Phalaenopsis violacea*. Many breeders name orchids after their relatives or friends, and one may be able to identify the origin of the hybrid by its name. Each hybrid name is limited to three words. When a hybrid is made between two genera, a bi-generic name is coined to identify the parent genera, thus *Arachnis* x *Vanda* = *Aranda*; *Arachnis* x *Renanthera* = *Aranthera*; *Arachnis* x *Phalaenopsis* = *Arachnopsis*. There are now more than 200 bi-generic names. When more genera are combined, the conjoined, polysyllabic name becomes quite a mouthful, for example, *Brassolaeliocattleya*. To prevent hundreds of such names appearing in orchid nomenclature, a family name ending in *ara* is given to polygeneric hybrids. This name is usually derived from the surname of the hybridiser or some famous orchidist whom he holds in high regard. Thus, *Arachnis* x *Renanthera* x *Vanda* (first produced by crossing an *Aranda* with *Renanthera*) was called *Holttumara* after Professor Eric Holttum. Add

Vandopsis to *Holttumara* and the combination has a new generic name, *Teohara*; add *Phalaenopsis* (or rather a *Paraphalaenopsis*) to *Holttumara* and we have *Bokchoonara*. There are now approximately 300 hybrid genera, a third of which contain more than two generic combinations.

By convention, in listing a hybrid, the pod parent is named first, followed by the name of the male parent which contributes the pollen. Thus, the statement *Vanda* Rothschildiana = *Vanda sanderiana* x *Vanda coerulea* means that the pollen was taken from *Vanda coerulea* and placed on the stigma of *Vanda sanderiana* when this hybrid was made. Reverse crosses do bear the same hybrid name, but the hybridiser would have reported his crossing as *Vanda coerulea* x *Vanda sanderiana*. The correct sequence is useful to the hybridiser because some plants cannot give fertile pollen.

Natural Hybrids

When the famous 19th-century British taxonomist John Lindley learnt that orchid hybrids had been produced by crossing different species, he suspected that natural hybrids must also occur in nature. Among the first to be recognised was *Phalaenopsis intermedia*, which was introduced into England in 1852. John Lindley noted that it was likely to be a hybrid between *Phalaenopsis amabilis* and *Phalaenopsis rosea*. It agreed with the former in foliage and in the tendrils of its lip; with the latter in its colour, in the acuteness of its petals and in the peculiar form of the midlobe of the lip. The flowers were also halfway in size between *Phal. amabilis* and *Phal. rosea*. The fact of this was proven by John Seden, who crossed *Phal. amabilis* and *Phal. rosea* and produced a plant which was identical to the jungle forms of *Phal. intermedia*. Interspecific crosses are particularly common with *Phalaenopsis*: *Phal. velentinii* (*violacea* x *cornu-cervi*), *Phal. gersenii*

(*violacea* x *sumatrana*), *Phal. amboinensis*, *Phal. singulifora* (*sumatrana* x *violacea*) and a natural hybrid between *Phal. kunstleri* and *Phal. sumatrana* being commonly found in Malaysia and Sumatra, while in the Philippines there are *Phal. leuorrhoda* (*aphrodite* x *schilleriana*), *Phal. vietchiana* (*equestris* x *schilleriana*) and *Phal. virataii* (a reciprocal cross of *Phal. vietchiana*).

Although they were originally believed to be sterile, most of these natural hybrids were in fact not mules but were highly fertile, and the frequency of back-crossing to either parent was only limited by the promiscuity of the insect pollinator. Indeed, having shown its willingness to visit either parent of the hybrid, there is no reason why the insect should become selective in its choice of nectar. In any large collection of Malaysian *Phalaenopsis violacea*, one can commonly find flowers which have stippling and barring closely resembling *Phalaenopsis valentinii*; the variability in the natural species is undoubtedly due to the inclusion of secondary natural hybrids produced in nature by introgressive hybridisation and back-crossing to *Phal. violacea*. Since hybrids have habitat requirements which are intermediate between the parents, and since both *P. violacea* and *Phal. valentinii* prefer shade whilst *Phal. cornu-cervi* tends to grow in exposed locations, back-crosses to *Phal. violacea* would be favoured. Back-crosses to *Phal. cornu-cervi* would be less competitive than the original *Phal. cornu-cervi* in the exposed locations.

The most famous natural hybrid from Singapore is *Vanda* Miss Joaquim (a cross between *V. hookeriana* and *V. teres*), which arose in the garden of Miss Agnes Joaquim in 1893. The equally famous *Arachnis* Maggie Oei is only the man-made counterpart of *Arachnis maingayi*, a natural hybrid between *Arach. hookeriana* and *Arach. flos-aeries*. *Arachnis* The Gem is another natural hybrid of *Arach. hookeriana*. The once popular and extremely variable *Dendrobium superbiens* is most likely a natural hybrid between members of the *Phalaenanthe* and *Ceratobium* groups of *Dendrobium*. Observing the wide variations in the wild forms of *Den. lineale*, *Den. mirbelianum* and *Den. warianum* collected in Papua New Guinea, Andree Millar concluded that cross-pollination among the three species was a continuous process, with back-crossing in all directions. Holttum suspected that *Paphiopedilum godefroyae* might be a natural hybrid between *Paph. bellatulum* and *Paph. concolor*, while Kamemoto and Sagarik suspected a hybrid origin for the *Paphiopedilum niveum* var. Ang Thong.

Inheritance of Morphological Characteristics

In morphological characteristics — size, shape, number, substance and colour of the flowers — orchid hybrids tend to be intermediate between the parents. This is especially true of interspecific and simple intergeneric hybrids. In interspecific hybrids (resulting from the selfing or the crossing of two members of the same species), the manifest result (phenotype) is influenced by the distribution of dominant and recessive traits.

With regard to the size and number of the flowers, both these characteristics follow the Law of Geometric Averages of MacArthur and Butler (1938), who first demonstrated its validity on tomatoes. According to this law, the size of the hybrid flower is the square root of the multiple of the flower sizes of the two parents. Similarly, the number of flowers of the hybrid is the geometric mean of the two

Clockwise from Top Left: *Phal. bellina* (syn. *Phal. violacea*, Borneo strain); *Phal.* Mok Choi Yew; *Phal.* Teoh Tee Teong (Mok Choi Yew x *violacea*)

parents. Thus, if one parent has six flowers each 10 cm across and the other parent has 50 flowers each 1 cm across, the hybrid will have, on average, 17 flowers each measuring 3.2 cm across.

Mehlquist (1946) showed that this law applied to orchids. When *Cymbidium eburneum,* which had single flowers, was crossed with *Cym. lowianum* (which had about 25 flowers), the resulting hybrids had 3–8 flowers per inflorescence, the average approaching the geometric mean of five instead of the arithmetic mean of 12. This was true for crosses of *Cym. eburneum* and *Cym. insigne, Cym. grandiflora* and *C. tracyanum.* But the rule does not always apply, and when *Cym. eburneum* was crossed with *Cym. devonianum,* a small-flowered species with 18–34 flowers, the resulting hybrid, *Cym.* Jean Brummitt, had the same number of flowers as *Cym. devonianum.*

Coming to tropical lowland orchids, the application of this rule is most obvious in the intersectional *Phalaenopsis* hybrids between large, floriferous whites and small, single-flowered species; and in this case, even the polyploidy of the white parent has been unable to overcome the limitation in the number of blooms on the spray. The large yellow strap-leaf *Vanda* inherit the yellow from *V. dearei,* which has 3–5 flowers, and hence the hybrids all have either short sprays and clear yellow flowers or long sprays but tessellated flowers from *V. sanderiana.*

When one of the parents is polyploid, the hybrid takes after the polyploid parent. Dominance is also exhibited by diploid plants which produce non-reduced or polyploid gametes. If the parents are equal in ploidy, the pod-bearing parent exerts a stronger influence in the hybrid, this coming from the cytoplasmic RNA of the ovum. (In the case of humans, that's about 3 percent.) That is one reason why the general preference is

to pollinate an *Eu-Phalaenopsis* with pollen from the star-shaped species, not the other way round.

Colour Inheritance

The control of colour inheritance in orchids was first discussed in *Cattleya* by the British geneticist C.C. Hurst, who was much assisted by the rediscovery of Mendel's paper on the inheritance of dominant and recessive traits in 1900. Hunt found that when true albinos were crossed with other true albino species, they did not always produce true albinos, as might be expected, but instead commonly produced all coloured reversions, both albinos and coloureds, or all albinos. In a paper published in 1913, Hurst concluded that colour in *Cattleya* was determined by the presence of two factors which he designated as C and R. C refers to the chromogen, a colourless colour precursor, and R to the enzyme which converts the chromogen into the coloured pigment. When both chromogen and enzyme are present in a plant, the flowers are coloured; when either or both are absent, the plant manifests as an albino. The genes which control the inheritance of these factors (and in fact any genetic trait) appear in pairs. If we refer to the active factor by capitals and a recessive factor (even in its absence) by small letters, the coloured plants may be designated as CCRR, CCRr, CcRR or CcRr. This designation implies that the colour-determining genes are dominant, and such is indeed the case. The albinos may be ccRR, ccrr or CCrr if they are homozygous, or ccRr or Ccrr if they are hybrid and heterozygous.

Hunt concluded that, in nature, albino *Cattleya* were either ccRR or CCrr, and he was able to categorise the various clones of 10 species into these two groups. The list has been expanded by Curtis and Duncan (1942), Mehlquist (1958) and Northen (1970).

Table 1. Result of Selfing Albino *Cattleya*

Group 1 ccRR		Group2 CCrr	
ccRR	albino parent	CCrr	albino parent
cR	gametes	Cr	gametes
selfed		selfed	
ccRR	progeny all albino	CCrr	progeny all albino

When crosses are made within the same group, whether by the selfing of a species or by interspecific hybridisation, the same genetic constitution, for example, albinism, is reproduced, and the progeny are all albino (Table 1). When a member of Group 1 is crossed with a member of Group 2, all the progeny will have a CcRr genetic constitution, and they will all be coloured (Table 2). If this heterozygous F1 hybrid (CcRr) is now selfed, we may expect seven white to every nine coloured siblings in the F2 generation (Table 3). Reselfing of any albino plant in the F2 generation produces 100 percent white hybrids (Table 4). However, if sibling crosses of the F2 generation are made, coloured forms will reappear in the F3 generation, unless the selected siblings have the same recessive gene (that is, either both parents are ccRR or both parents are CCrr) or one or both parents are doubly recessive (ccrr).

In attempting to produce white *Cattleya,* we have several problems in applying the knowledge above:

1. The classification into the two groups ccRR and Ccrr refers only to the specific clones identified by the various authors.
2. Tinged albinos must not be confused with true albinos because tinged albinos behave in breeding as coloured plants.
3. Polyploidy would introduce an entirely different set of sums.

Table 2. Result of Crossing Albino *Cattleya* from Two Groups

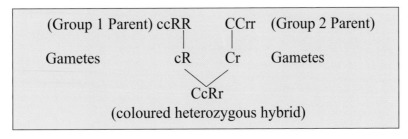

(Group 1 Parent) ccRR   CCrr   (Group 2 Parent)

Gametes   cR   Cr   Gametes

CcRr
(coloured heterozygous hybrid)

All the second-generation *Cattleya* produced by selfing retain the same hybrid name. Only one of the seven albino *Cattleya* is a double recessive and will be bred true if it is selfed.

Table 3. Results of Selfing the First-Generation Coloured Heterozygous Hybrid Cattleya

Gametes	CR	Cr	cR	cr	
CR	CCRR	CCRr	CcRR	CcRr	4 coloured
Cr	CCRr	CCrr	CcRr	Ccrr	2 coloured, 2 albino
cR	CcRR	CcRr	ccRR	ccRr	2 coloured, 2 albino
cr	CcRr	Ccrr	ccRr	ccrr	1 coloured, 3 albino
Total					9 coloured, 7 albino

All the second-generation *Cattleya* produced by such a crossing will still have the same hybrid name as the first-generation hybrid. Only one of the seven albino *Cattleya* is a double recessive.

Table 4. The Identification of a Double Recessive Albino Hybrid

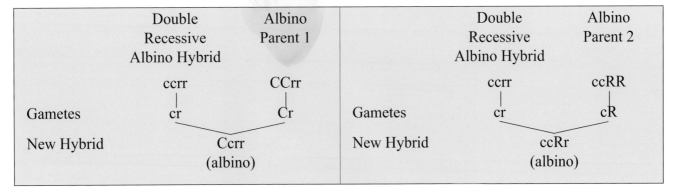

Double Recessive Albino Hybrid	Albino Parent 1		Double Recessive Albino Hybrid	Albino Parent 2
ccrr	CCrr		ccrr	ccRR

Gametes   cr   Cr   Gametes   cr   cR

New Hybrid   Ccrr   New Hybrid   ccRr
                        (albino)   (albino)

The method uses a back-crossing to each of the original two parent species.
In both instances, all the progeny are white.

Above: *V.* Rose Davies

Right, Top and Bottom: Two quite different clones of *V.* Overseas Union Bank (*V.* Mimi Palmer x *V.* Kasem's Delight). Blue is a much sought after colour in orchids, as much today as 25 years ago when the *V.* Rose Davies was photographed. Tessellation is another characteristic imparted by *V. coerulea* to its hybrids. *V.* Mimi Palmer (JVB x *tessellata*) has bluish grey tessellations over a grey, small-sized flower that is strongly scented. Both colour and scent have come through in this large-flowered hybrid, *V.* Overseas Union Bank.

Opposite: *Lc.* Longriver Compton FCC/AOS

With regards to the other genera, the inheritance of whites have either not been properly worked out or not been announced. Thus, the safest way to ensure an all-white generation is to self the plant. Again, beware the tinged albinos. True albinos have no coloured pigments on the plants except for green and, occasionally, yellow; the undersides of the leaves and the root tips are green.

Colour in flowers is determined by three types of pigments: anthocyanins and anthoxanthins, which are water soluble, and plastid pigments, which are insoluble in the cell sap. The anthocyanins are responsible for red, blue and magenta colours, and in *Cattleya* they are controlled by the presence of the dominant genes C for chromogen and R for the enzyme. Because the genes are completely dominant, homozygous plants with CCRR genotype do not produce darker flowers than the heterozygous plants with CcRR, CcRr or CCRr constitution. There is speculation that other genes may also be responsible for colour, determining intensity and distribution, but these have not been worked out for orchids, except for the inheritance of the semi-*alba* characteristic, to which we shall come later.

The anthoxanthins produce a colour range from pale ivory to deep yellow, and the colour is intensified by a rising pH. W.B. Storey postulated that they are controlled by multiple genes in *Cattleya*. According to Rebecca Northen, the purple colour is dominant over the yellow colour in *Cattleya*, while in *Laelia* the yellow is dominant. In *Laeliocattleya*, yellow is dominant over purple. Thus, when a purple *Cattleya gigas* crossed with a yellow *Cattleya* Hardyana is selfed, a quarter of the plants in the F2 generation will be yellow and they will breed true for yellow. Furthermore, crossing this yellow plant back to the yellow *Cattleya dowiana* will not produce all yellow offspring.

A crossing between the purple *Cattleya* and the yellow *Laelia* will give all yellow flowers in the F1 generation, because the *Laelia* yellow is dominant. By selfing the F1 generation, it is possible to produce a homozygous yellow *Laeliocattleya* with a full *Cattleya* form, and this plant can then carry the dominant yellow gene to the next generation of yellow *Laeliocattleya*, which will have even better shape (dominant yellow *Laeliocattleya* x *Cattleya*).

It is not quite so simple in other genera. In *Vanda*, for instance, many musty colours result when unique colour combinations are tried — blue with yellow or blue with brown.

Coloured Lips

The breeding of bicoloured, semi-*alba* or white with coloured-lip flowers has given us some of the most attractive hybrids in *Cattleya*, *Dendrobium* and *Phalaenopsis*. Mehlquist conceived the ingenious theory that colour distribution in flowers was determined by special genes called PP, Pp or pp, which were inherited independently of the CC and RR genes. If the CR phenotype is homozygous recessive ccrr, the PP genes are inactive and the entire flower is white. When both the C and R genes are present in a plant, either PP or Pp will cause the plant to be coloured (P is a dominant gene). The homozygous, recessive genotype pp, in the presence of both C and R, causes the plant to produce a coloured lip, but no colour appears on the petals and sepals (i.e. the plant is white, with a coloured lip). The selfing of a semi-*alba* would produce either semi-*albas* or whites, or a combination of both, but no solid-coloured flowers. On the other hand, crossing white to semi-*alba* may produce whites and semi-*albas* or coloureds, whites and semi-*albas*, depending upon the phenotype of the white parent.

The Premature Fading of Flowers

Premature fading is common in some orchids, for instance, in *Phalaenopsis* which combine *Eu-Phalaenopsis* with members of the section *Stauroglottis*, in *Renanthera*, *Aranthera* and *Renanstylis*. In *Phalaenopsis*, the problem can only be avoided by selecting the *Stauroglottis* parent that is colour fast: we have that in Roy Fields' strain of *Phal. luddemanniana* var. *ochracea* (later identified as a form of *Phal. fasciata*). In the other hybrids, the problem lies with the *Renanthera* parent which has a large, branching inflorescence and flowers that open gradually over an extended period. By the time the inflorescence is two-thirds open, the earliest flowers begin to fade. However, there are a few clones of *Ren. storiei* that maintain an even colour intensity throughout the inflorescence, a quality that is transmitted to their progeny. By using such clones exclusively, Asian hybridisers have now produced a good cluster of red orchids that do not express premature fading.

Gene Linkages

The characteristics that we have described above obey Mendel's Second Law of Inheritance, or the Law of Independent Assortment, which states that genetic traits are inherited independently of each other. This law applies only to certain traits. When two characteristics are determined by genes which are located practically next to each other on the same chromosome, they may be inseparable, and the two traits are then inherited together or not at all.

The Cytogenetics of Hybridisation

In order to continue the discussion of hybridisation, it is now necessary to describe the basic elementary concepts of cytogenetics.

Genes which are responsible for hereditary characteristics — the appearance and function of every living cell and organism — are carried on specialised structures within the cell nucleus known as chromosomes. Each gene has a specific locus on a particular chromosome and a single, specific function. The chromosomes become visible only during a brief period of cell division, and they are most obvious during metaphase. The vast majority of orchid plants, like all other living creatures, have two sets of chromosomes and they have the cytological designation of 2x, where x refers to the basic number of chromosomes, in a complete set. In cytogenetic terms, these normal plants are referred to as diploid. Some plants have three or more sets of chromosomes, and they are called triploid (3x), tetraploid (4x), pentaploid (5x), hexaploid (6x), etc., depending on the number of sets of chromosomes that are present in their somatic or body cells. These plants are also referred to as polyploids. Polyploidy is not confined to orchids. In fact, most important crops are polyploid: alfafa, potato, coffee and peanut are tetraploid, sweet potato is hexaploid, and the common banana is triploid.

Just before the cell reproduces itself, the chromosomes are duplicated, and the newly duplicated pairs gather in the centre of the cell and are held by two spindles. Quite suddenly, the spindles pull the chromosomes to opposite ends of the cell, and a new cell wall and nuclear membrane appear, resulting in two cells with the same chromosome number and the exact constitution of the original cell. This process is called mitosis.

During the production of sex cells or gametes (which become pollen and ovules), the chromosomes come together in pairs. The corresponding members of each pair now travel to opposite ends of the nucleus. When the cell divides, each of the new cells or gametes contains half the number of chromosomes which were present in the original cell. The process is called meiosis. The number of chromosomes in the gamete is written as n. In diploid cells n = 2x. However, there is considerable confusion in the usage of symbols and it is common practice to use x and n synonymously.

Examination of a chromosome at metaphase shows that it consists of two short arms and two long arms, with a constriction or centromere between the short and long arms. During meiosis, the chromosomes come together in pairs, and the arms of paired chromosomes may break and exchange parts with each other. This process, called crossing over, results in new hereditary combinations in the reconstituted chromosomes and in genetically unique gametes.

Chromosome Numbers

Orchids which belong to a natural genus or a section of a genus have a constant chromosome number except for those few individuals which are polyploid. Sometimes, the entire genus and even larger taxonomic groupings have the same chromosome number because of their common ancestry. The commonest chromosome number is 38, the typical number in vandaceous orchids and in *Dendrobium*. There is a very wide range in chromosome numbers in the genus *Oncidium*, with *Oncidium pulsillum* having only 10 chromosomes while *Oncidium varicosum* has 168. Within the genus *Paphiopedilum*, there are subgroups with 26, 32, 36, 38 and 42 chromosomes.

The commonest method for determining the chromosome number of an orchid is to count the chromosomes in squash preparations of dividing cells from the root tip. These are somatic cells which have the full complement of chromosomes, generally with two sets,

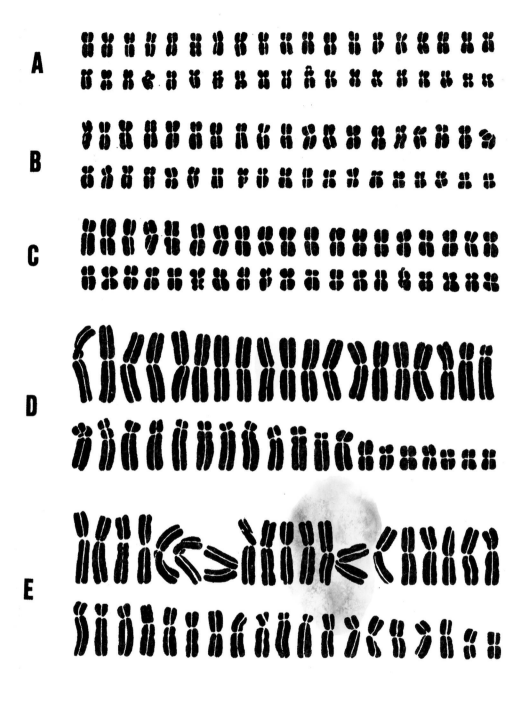

one derived from each parent. In the diploid *Vanda*, the root tip count would yield 38 chromosomes, and the basic number of each set (x) is 19. Thus, if the squash preparation shows a *Vanda* to have 57 chromosomes, it is triploid (3 x 19 = 57); if 76, tetraploid (4 x 19 = 76). Due to an error at meiosis, gametes may lose or acquire a few chromosomes, and the resultant zygote will show a deletion or addition of an odd chromosome. These plants whose chromosome numbers do not add up to exact multiples of the basic number are known as aneuploids, and they are usually sterile or subfertile.

When two species of plants in the same tribe have an identical chromosome number they are likely to be compatible in hybridisation. If the chromosome numbers are different, the plants are not likely to breed with each other even though they may belong to the same genus.

Chromosome Size and Morphology

Haruyuki Kamemoto found the most remarkable variation in chromosome size in the genus *Phalaenopsis*, while in the other genera within *Sarcanthus*, such as *Vanda*, *Asocentrum*, *Renanthera*, *Arachnis*, *Aerides* and *Angraecum*, the size of the chromosome is identical. Strange as it may seem, the large-flowered members of the *Eu-Phalaenopsis*, such as *Phalaenopsis sanderiana* and *Phalaenopsis schilleriana*, have the smallest chromosomes; *Phalaenopsis lindenii* and *Phalaenopsis equestris* of the *Stauroglottis* section have slightly larger chromosomes; *Phalaenopsis lueddemanniana* and *Phalaenopsis boxalii* still larger; whilst *Phalaenopsis manii*, *Phalaenopsis violacea* and *Doritis pulcherrima* have the largest chromosomes.

When a hybrid is produced with parents of different chromosome sizes, the chromosomes

Above: Chromosome Shape, Size and Numbers in Various Species of *Phalaenopsis* and in *Doritis*. A — *Phal. schilleriana*; B — *Phal. lindenii*; C — *Phal. lueddemanniana*; D — *Phal. violacea*; E — *Doritis pulcherrima*. (Diagram by courtesy of Prof. H. Kamemoto, from the proceedings of the Fourth World Orchid Conference, 1963. Malayan Orchid Society, Singapore.)

do not pair well at meiosis, rendering the plant almost sterile. The degree of relationship between parent species has the most profound effect on the fertility of the hybrid: the more distant the relationship, the fewer the chromosome pairs and the poorer the fertility.

If an interspecific hybrid shows perfect chromosome pairing at meiosis, one can expect it to be extremely fertile, and there will be normal recombination of characteristics in the second generation. Sometimes, even intergeneric hybrids show near perfect chromosome pairing. Thus, *Ascocenda* made from strap-leaf *Vanda* shows a good chromosome pairing with 18–19 bivalents, and this is also true for crosses between *Vanda sanderiana* (syn. *Euanthe sanderiana* as classified by Schlechter, 1915; Holttum, 1949) and strap-leaf *Vanda*. Proper chromosome pairing allows for repairs to be made to the chromosome and the elimination of defective genes. In the long term, this is essential for the survival of the lineage.

In the short term, if meiosis is slightly irregular in the hybrid, even if only one or two pairs fail to synapse, there will be diminished fertility and appearance of aneuploid progeny in the next generation.

If the chromosomes of the parent species in a hybrid are dissimilar, only a limited number of bivalents will be produced at meiosis and the plant's fertility is greatly impaired. This is the situation in semi-terete *Vanda*. Occasionally, such irregularity at meiosis results in the formation of unreduced 2x and 4x gametes, in addition to those with unbalanced chromosome numbers. This behaviour has been observed whenever diploid semi-terete *Vanda* has been successfully crossed with another diploid plant, and the results of the crossings are a generation of triploids and pentaploids in varying proportions. The resultant hybrids retain the full chromosome sets from their semi-terete parent and are fairly homogenous.

Unreduced pollen also resulted in triploid progeny when *Dendrobium phalaenopsis* Lyon's Light No. 1 was crossed with *Dendrobium* Neo-Hawaii. Both these parent plants were diploid, and if we refer to the *Phalaenanthe* genome (the basic set of chromosomes) as P and the *Ceratobium* genome as C, *Dendrobium phalaenopsis* can be written as PP and *Dendrobium* Neo-Hawaii as PC. When *Den*. Neo-Hawaii pollen was transferred to *Dendrobium phalaenopsis* Lyon's Light No. 1, the progeny was PPC; the other way around it was 2.5P 0.5C; *Dendrobium* Neo-Hawaii, selfed, was 1.5P 1.5C; and of course *Dendrobium phalaenopsis*, selfed, was PPP. In the hybrid *Dendrobium* (Neo-Hawaii x *phalaenopsis*), the *Dendrobium phalaenopsis* effect would be more marked when *Dendrobium phalaenopsis* was used as the pollen parent than the other way around. Here, the dominance from the genomic effect overrides the smaller contribution of cytoplasmic RNA.

Autotetraploids have been identified in *Aerides odoratum* var. *immaculatum*, *Cattleya labiata*, *Cymbidium pumilum*, *Dendrobium kingianum*, *Dendrobium nobile*, *Dendrobium phalaenopsis*, *Doritis pulcherrima* var. *buysonniana*, *Epidendrum atropurpureum*, *Phalaenopsis schillerana*, *Rhynchostylis gigantea*, *Vanda concolor*, *Vanda denisoniana* and *Vanda tricolor*. Hexaploids were found in *Dendrobium kingianum*, *Renanthera coccinea* and *Vanda spathulata*. Octoploids were found in *Phalaenopsis amabilis*. Triploids occur in *Dendrobium biggibum* and *Dendrobium nobile*. Diploid and tetraploid populations often have distinctive distribution, for example, the diploid *Doritis pulcherrima* is found in Peninsular Thailand and Malaysia, while the tetraploid *Doritis pulcherrima* var. *buysonniana* exists in northern Thailand. In Australia, Keith Maxwell found that diploid and tetraploid

Dendrobium kingianum also had distinctive distribution which was related to altitude and exposure. Triploids occurred where the two types were in contact.

Allopolyploids which have multiple sets of chromosomes from two or more species or genera may arise in several ways:

1. By spontaneous chromosome doubling in meristematic tissue
2. By encouraging chromosome doubling through the use of colchicines on protocorms
3. By the mating of unreduced gametes
4. Through hybridisation of autotetraploids with diploid, autotetraploid or allotetraploid plants

The allotetraploids are the most valuable plants for the development of intersectional and intergeneric hybrids. They have been responsible for the major breakthroughs in breeding semi-terete and quarter-terete *Vanda* (among them *V.* Tan Chay Yan and *V.* Nellie Morley), second-generation *Aranda* and a beautiful range of large, overlapping, heavy-textured *Phalaenopsis*; and they will soon be leading the development in intermediate type, warm-growing *Dendrobium*, and in *Rhynchovanda*.

Allotetraploids have been artificially induced by treating protocorms and other meristematic tissue with colchicine. As early as 1963, Mrs. Menninger reported that she was able to produce an allotetraploid

Opposite, Clockwise from Top Left: **Tetraploid** *Den. phalaenopsis*-type; tetraploid *Den. phalaenopsis* (Photos by courtesy of Prof. H. Kamemoto); *Den.* Burana Pearl — a fine hybrid like this one would not have been possible without a knowledge and application of polyploidy.

sectional chimera in *Cymbidium* Coningsbyam 'Brockhurst' by treating the 'eyes' of dormant back bulbs with colchicine solution. After piercing the bud with a fine needle, she soaked it in 0.3 percent colchicine solution for 74 minutes, and 10 days later she again soaked the bud in colchicine, this time in a 1 percent solution. A root tip study two years later showed that the new growth had 80 chromosomes, i.e. it was tetraploid. Subsequently, Gavino Rotor Jr. was able to obtain viable seed from a sterile, triploid *Cattleya* Mary Schroeder after treating the flower buds with colchicine and by using pollen from the treated flowers. It was presumed that the treatment had caused a reduplication of the chromosomes. (The process had long been known to medical scientists who observed the phenomenon when human white blood cells in short term culture are subjected to colchicine to produce mitotic arrest for the purpose of karyotyping. Albert Levan called it endo-reduplication. This tetraploidy was a laboratory aberration that was not compatible with normal human life.)

In 1996, Donald Wimber and Ann Van Cott reported that exposure of protocorms from seeds and from meristem explants of diploid *Cymbidium* hybrids to 0.05 percent colchicine for 10–21 days caused 40 percent of the treated seedlings and 43 percent of the meristem plantlets to become tetraploid.

The appearance of allotetraploid and aneuploid plants in meristem seedlings is now fairly commonplace, and great advances in hybridisation may be expected when the full potential of this technique is exploited.

Among *Cymbidium*, *Cattleya* and

Paphiopedilum, there have been numerous outstanding tetraploids. The first to blaze the trail were *Cymbidium* Alexanderi 'Westonbrit', *Cymbidium* Pauwelsii 'Compte d'Hemptinne', *Cattleya* Enid, *Cattleya* Titrianaei, *Cattleya* Bow Bells and *Cattleya* Joyce Hannington. Professor Mehlquist found that the first cross in *Paphiopedilum*, *Paph.* Harrisianum, yielded one tetraploid with 64 chromosomes. (However, the genus *Paphiopedilum* is interesting from the breeding point of view because at least two great parents were not tetraploids but were, in fact, fertile triploids — *Paph. insigne* 'Harefield Hall' and *Paph.* F.C. Puddle FCC/ RHS.) The famous allotetraploids which are of greater relevance to the tropical grower are *V.* Josephine van Brero, *V.* Emma van Deventer, *V.* Miss Joaquim, *Phalaenopsis* Doris and a whole range of large, white *Phalaenopsis*, the pink *Phal.* Zada, *Arachnis* Maggie Oei 'Red Ribbon', *Rhynchovanda* Sagarik Wine, *Dendrobium* Diamond Head Beauty and many of its award-winning progeny, such as *Dendrobium* Macrobig, *Dendrobium* Jaquelyn Thomas UH44 (white) and UH232 (lavender). Many of those plants were selected by growers and hybridisers who had no knowledge of their chromosome make-up. They were outstanding plants because of the quality of their blooms, and as parents they were extraordinary because of the tremendous quality of their hybrids.

The Characteristics of Polyploid Plants

The common effect of polyploidy is to increase the vegetative portions of the plants, making them more lush and more vigorous than the corresponding diploids when they are fully grown. As seedlings, though, they may be rather slow-growing if their ploidy is 4x or above, but triploids are fast-growing, spurting ahead of all the other varieties. Frequently, polyploids produce larger flowers with overlapping petals and sepals which are of heavy substance. Nevertheless, the effect is not universal and many polyploids have small flowers with narrow floral parts, and one tetraploid, *Dendrobium* Jaquelyn Thomas, is known to drop buds. Under the microscope, the polyploid cells are larger.

Kamemoto reported that in the *Den. phalaenopsis* and intermediate groups of warm-growing *Dendrobium*, the flowers of tetraploids or amphidiploids were slightly fuller than the diploids, and the flower spikes were sturdier and more erect. The individual flowers of the triploids were superior to either the diploids or tetraploids. Practically all the award winning *Dendrobium* in Hawaii are polyploid or aneuploid above the triploid level. Among the cool-growing *Dendrobium*, Jiro Yamamoto observed that the tetraploids were healthy, fast growers which were resistant to insects and disease, and most of them produced larger blooms. (The polyploid Yamamoto *Dendrobium* are hailed throughout the world as a class of their own, being round, thick-textured, brightly coloured, floriferous and free-flowering.) However, the majority of tetraploids were late bloomers and some were poor in providing blooms. The autotetraploid *Rhynchostylis gigantea* 'Sagarik strain' was difficult to grow, but its triploid progeny are extremely vigorous and free-flowering. Allotetraploid semi-terete *Vanda* are tough but have been surpassed by their triploid progeny in vigour, resistance to disease, floriferousness, and the shape of the flowers have been improved by crossing back to round strap-leaf *Vanda*. Triploids and pentaploids, are characterised by aberrant chromosome behaviour at meiosis and a high degree of sterility. Thus, they are best reproduced by vegetative propagation. Meiosis in triploids is more irregular than in pentaploids, and the latter is often fertile.

Offspring which result from the occasional triploid crossing show a wide range of chromosome numbers with a high percentage of aneuploids ranging from 2x to 5x, while those which result from the pentaploids range from the 3x to the 4x level. The pentaploid (5n) parent contributes either two (2n) or three (3n) full sets of chromosomes.

Allotetraploidy or amphidiploidy restores fertility to sterile diploid hybrids between distantly related orchids. Most of the major breakthroughs in orchid hybridisation were due to the chance discovery of amphidiploid stud plants. When amphidiploids are bred with ordinary diploids, the resultant triploid progeny are vigorous and free-flowering. They will exhibit the prepotency of the amphidiploid parent because (1) it has contributed twice the number of chromosomes, and (2) entire sets of chromosomes (genomes) are tranferred from the polypoid parent without genetic arrangement at meiosis. The best examples of highly successful amphidiploid-diploid crossings are those made with *V.* Josephine van Brero, *V.* Emma van Deventer, *V.* Dawn Nishimura, *Ascda.* Navy Blue and *Ascda.* Yip Sum Wah; with tetraploid-diploid, *V. denisonniana, Rhynchostylis gigantea* and *Aerides lawrenceae.*

Two outstanding polyploid *Dendrobium*

Above, Left: Amphidiploid *Den.* Macrobig is a Hawaiian hybrid between two very dissimilar sections of *Dendrobium*. It opens an avenue for breeding with the fascinating *Latourea* section. (Photo by courtesy of Prof. H. Kamemoto)

Above, Right: Tetraploidy has been successfully engineered in this *Dendrobium phalaenopsis*-type hybrid at the Singapore Botanic Gardens. Although the petals are not overlapping, the substance is tremendous and the colour is very deep. The flower is quite large.

Opposite: *Den.* (Ryzhkova x *lasianthera*), 2n and 4n. Dr. Tim Yam created the tetraploid plant at the Singapore Botanic Gardens. He points out that the tetraploid flower is larger, broader, stiffer, its petals are more erect, but the colour is lighter in the tetraploid flower although both have the same mix of genes. Dr. Yam is holding the tetraploid flower in this picture.

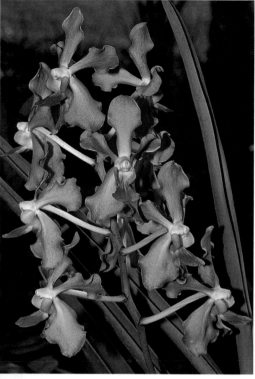

Four famous hybrids of the tetraploid *Vanda* Josephine van Brero (*V.* JVB)

From Left to Right:

Renantanda Charlie Mason var. Seng Heng AM/MOS (1964) (left) is one of the loveliest hybrids bred from *V.* JVB. Arrangement of the flowers is excellent and colour is good, but there is that little bit of premature fading. It is worth repeating the cross given the superior forms of contemporary *Renanthera*.

Renantanda Ammani (above) has stronger colours but fewer and poorer-shape flowers than *Rntda.* Charlie Mason. One clone did win an Award of Merit.

Vanda Chia Kay Heng (JVB x Dawn Nishimura) (opposite, left). Both parents are tetraploid. Hence, its blue colour is more intense than the next hybrid where a diploid blue has to cope with the tetraploid brick-red of JVB.

Vanda Tan Chin Tuan (JVB x Rothschildiana) (opposite, right) was the best blue JVB-type of the 1950s and 1960s. The blue hue is very delicate. Worth repeating if a 4n *V.* Rothschildiana is available.

During the past two decades, knowledge of ploidy has been translated into practice in Thailand and Singapore. An astute breeder is usually able to spot the tetraploid plant, but when he cannot be certain, he will perform a chromosome count. Tetraploidy has also been engineered in the laboratory.

Freedom of Flowering

Freedom of flowering in orchid plants appears to be affected by (1) polyploidy, (2) the flowering characteristics of the parents, and (3) their behaviour as seedlings. Within a crossing, it may vary greatly from clone to clone.

Triploidy invariably produces floriferousness. The triploid hybrids of *V.* Josephine van Brero and *V.* Emma van Deventer, the first-generation *Aranda* produced by breeding with *V.* Dawn Nishimura, and the intergeneric triploid hybrids with *Rhynchostylis* gigantea 'Sagarik strain' 4n are all continuous bloomers. Tetraploids tend to be seasonal, with the exception of *Arachnis* Maggie Oei 'Red Ribbon'.

When a hybrid is produced from a species which is strictly seasonal, it may inherit this characteristic (such as in crosses with *Arachnis flos-aeries insignis* and *Arachnis breviscapa*), but sometimes a breakthrough is achieved and the hybrid flowers continuously. Some species, although seasonal, produce hybrids which are continuously blooming, such as *Ascocentrum miniatum*, *Rhynchostylis gigantea* and *Aerides lawrenceae*. Again, a few species which normally cannot flower without a low temperature stimulus, such as *Ascocentrum curvifolium*, give hybrids which are extremely floriferous over a wide temperature range. Occasionally, it has been possible to extend the flowering season of the hybrid by making crosses between two species which flower at different times of the year. The hybrid will flower during the flowering season of each

parent and also in between. *Cattleya* was the first to be bred in this way.

The intermediate-type *Dendrobium* have partially broken through the seasonal flowering traits of the *Phalaenanthe* and *Ceratobium*, flowering throughout the year, but they still have peak flowering seasons possibly during the flowering seasons of their parents. When back-crossed to the *Phalaenanthe*, they become seasonal again.

Holttum introduced a rule, now applied by all growers in Singapore as a basic for selecting plants for cut-flower purposes, that "the early flowering in a seedling is an indication of the freedom of flowering for the rest of the life of the plant." In the striking case of *Arachnis* Ishbel (*A. maingaya* x *A. hookeriana*), a dozen seedlings flowered at the Singapore Botanic Gardens, but only the first to flower proved to be really free-flowering. It also applied to *Aranthera* Mohamed Haniff.

A complete picture of the genetics which determine the flowering of orchids has not emerged, and today hybridisers still go by trial and error. Some outstanding parents have been identified. Thus, when *Arachnis* Maggie Oei (*hookeriana* x *flos-aeries*) was first made with the common form of *Arach. hookeriana*, it refused to flower, but when the same crossing was repeated with *Arach. hookeriana* var. *luteola*,

Opposite, Top Left and Right: *Vanda* Miss Joaquim. The smaller flower is the original diploid variety bred by Miss Agnes Joaquim, for which Holttum had proposed the varietal name Agnes. The larger, fuller flower is the Hawaiian tetraploid variety, Douglas.

Opposite, Bottom: *Vanda* Nellie Morley FCC/AOS

Left: *Vanda* Tan Chay Yan var. Tan Yeow Cheng AM/MOS (1959)

the hybrid was fantastically free-flowering. *Arach. hookeriana* var. *luteola* went on to produce the free-flowering *Aranthera* James Storie which the common *Arach. hookeriana* was unable to do. Modern first-generation *Aranda* derived from *Arach. hookeriana* are made with the variety *luteola*. But the common *Arach. hookeriana* itself is extremely floriferous, which is astonishing. Even more strange is that when the Singapore Botanic Gardens crossed it with the equally free-flowering *Vanda hookeriana*, all the seedlings refused to bloom.

When two orchids with different requirements for flowering (such as low temperature and dry season vs. tropical heat and year-round rainfall) are bred, the resultant hybrid may be able to adapt to a wide range of growing conditions. When Singapore's *Vanda* Miss Joaquim won her First Class Certificate (FCC) and a Cultural Commendation from the Royal Horticultural Society on June 15, 1897, the plant had flowered in the garden of Sir Trevor Lawrence in Burford, Dorking, England. The inflorescence carried 12 flowers and 3 buds, which is approximately twice the maximum number of blooms that we see in Singapore. The critique in the *Gardeners' Chronicle* said: "What a beautiful hybrid it is. The flowers closely resemble *V. hookeriana* — which is the more fortunate because scarcely anyone succeeds in growing that parent well — while it seems as easily grown as *V. teres*."

Holttum asked for the cross of *Spathoglottis plicata* with *S. aurea* to be made at Penang Hill because *S. aurea* would not flower in the lowlands. The result hybrid, *S.* Primrose is free-flowering in full sun in Singapore. Thai hybridisers have produced an enormous number of superb hybrids by using pollen from species that only do well in Chiangmai (*Rhyncostylis gigantea*, *V. coerulea* and *V. denisonniana*) on their breeding stock in Bangkok.

The Importance of the Pedigree

Phenotype refers to the expressed characteristics of the genes in an individual; genotype to the individual's true genetic makeup.

There must be some good points in an orchid to inspire any person to breed with it, but experienced breeders know that it is always important to check on that orchid's pedigree. Do not rely on phenotype alone. Check the genotype. In breeding *Dendrobium*, for instance, if one is breeding for roundness, the *Dendrobium* must have a very high percentage of *Den. phalaenopsis* in its makeup. If there is too much *Ceratobium*, the resultant hybrid will not be round. Roy Fukumura spoke of his experience with *Dendrobium* Ram Misra to illustrate this point. The *Dendrobium* had seemed like a good parent because it had good round *Phalaenopsis*-type shape and good colour. The next generation produced open-petalled flowers. When he checked the parentage of *Den.* Misra, he found that it had too much *Ceratobium*. In *Vanda*, without sufficient *V. sanderiana*, the resultant hybrid will not have overlapping form.

In *Phalaenopsis*, look for abundance and repetition of famous parents in the pedigree and also the petals of the white parent; it should overlap if the other parent is star-shaped. In the final analysis, most whites are essentially *Phal. amabilis*. One way to improve on an already good hybrid is to do a selfing (a self-pollination), or alternatively to cross one selected plant with another of the same hybrid and then to select the best plants from among the seedlings. The process can be repeated for several generations. This was what Roy Fields did with his *Phal.* Zada. However, when he reached the seventh generation, Fields found that the flowers became much smaller, though they had perfect shape, excellent substance and intense colour with spotting. It is interesting

On these two pages: *Vanda* Charles Goodfellow x *Ascda*. Nopporn Gold. The four pictures illustrate the different selected colour forms from a single breeding. The actual range is much wider, but these four clones have been selected by the breeder for further study. They are all beautiful. The choice would depend on one's preferences and what one wishes to develop, and some compromise may have to be made. For instance, the shape can be improved in the pure lemon green (*alba*) form and in the spotted flower.

to read that Fritz Hark, the German breeder whose *Phal.* Lipperose (Ruby Wells x Zada) became an important parent of modern pinks, followed a similar approach of line breeding to get the outstanding clones of *Phal.* Lipperose. Was he recollecting the genes of *Phal.* Zada? Selfing, or back-crossing to one parent, is the method employed by breeders of plants and animals to collect the good genes and to discard the poorer genes. When one is able to do this successfully, the percentage representation of a species in the parentage (e.g. *V. sanderiana*) becomes less important than the process of selection. When too much inbreeding results in loss of hybrid vigour, an outcrossing is necessary: afterwards, one may line breed again.

It is important to realise that when we say that a certain plant is a good parent, we are speaking of a specific plant (or its clone), and are not referring to all members of the species or the hybrid.

A careful hybridiser will spend months or years to trace the progeny of a particular plant that he wishes to use for hybridisation. When he is aware of the plant's strength as well as its weaknesses, he might be able to 'guess-timate' his chances of success before making a hybrid. But nothing is guaranteed. Syed Yusof Alsagoff is the foremost breeder of great orchids in Singapore. Over the past 40 years, he has spent an inordinate amount of time studying orchids and is aware of the good and bad points of every plant that he uses in breeding. He started by offering his services to be Secretary of the Award Committee of the Malayan Orchid Society, possibly the best place to start. One saw all the finest plants.

Selection

In any breeding, one generally expects a wide variation among the flowers of individual plants. One plant may have large flowers, but they may have poor form and colour. One may have excellent form but few flowers. There are endless possibilities, albeit fewer when a tetraploid parent is used. To have a good chance of obtaining a good plant, one needs to grow quite a few to the flowering stage. Most breeders keep 20–100 plants, but in Thailand, a farmer would often keep the entire crossing to select the best plant from it. That does not mean that the entire batch of seedlings from a sowing is grown to maturity. A batch is selected for accelerated nurturing, while the rest are kept back in a flask or in the compot before one decides what to do with them.

At the 11th World Orchid Conference in Miami in 1984, Treekul Sophonsiri of Thailand said that a beautiful hybrid must also meet the following additional requirements: early flowering, frequent free-flowering, easy and rapid growth. Modern Thai hybrids meet these criteria.

Care of the Breeding Stock

In an article in the *American Orchid Society Bulletin* (October 1993) wherein he imparted his experience of 50 years with orchids, Roy Fukumura, at 88, had this to say about the breeding stock:

> After a plant has set a seed pod (capsule) allow it to rest until it regains its full strength and is able to bear a healthy seed pod again. Do not set seed on the same plant often. Once I used a certain Ascda. *Yip Sum Wah every year for three consecutive years, and discovered that the second and third remakes were not as good as the original batch. Similarly, when I remade some* Vanda *and* Ascocenda *crosses using the same plants, I discovered this to be a bad practice.*
>
> Once the seed pod is set, do not move the plant, touch the seed pod, overwater, feed excessively or spray insecticides. These actions may cause the seed pod to be ruined. The environment around the plant is important too, and may help or hinder the proper development of the seed pod.

Comment

It is one thing to make a hybrid, but quite another creating one that will mark a milestone in the quest for excellence, or more importantly one that will lead to a greater thing. Some of the great breeders from Southeast Asia registered only one hybrid, but that single hybrid immortalised their name. The saga began with Miss Agnes Joaquim, who made her cross or discovered the hybrid of *V. teres* and *hookeriana* in her Tanjong Pagar garden in 1893. *Vanda* Miss Joaquim was still the pride of gardeners in this region well into the 1950s, and it is the parent of scores of beautiful hybrids. There was a long silence until 1936 when van Brero managed to cross *V. insignis* with *V. teres* in Tjipaganti, Java. The resultant *V.* Josephine van Brero led to a string of award-winning semi-terete *Vanda* that shared many distinct sibling similarities when it was crossed with *V. dearei*, *V. sanderiana*, *V. Rothschildiana*, *V. Pukele* and other large, round, colourful strap-leaf *Vanda*. Quite fortuitously, his countryman W. van Deventer also produced a tetraploid *Vanda* hybrid, *V.* Emma van Deventer, parent of the famous Hawaiian *V.* Nellie Morley.

Hybrids from *V.* Josephine van Brero and *V.* Emma van Deventer earned many awards from the various orchid societies during the 1950s and early 1960s, including an FCC from the RHS for *Vanda* Tan Chay Yan (*V.* JVB x *V. dearei*) bred by Tan Hoon Siang, and an FCC from the AOS for *V.* Nellie Morley (Emma van Deventer x *V. sanderiana*) bred by Morley. The pod-bearing plants were zealously guarded. *V.* JVB continued to be available only in Singapore for a while, but *V.* Emma van Deventer (4n) never left Hawaii.

John Laycock was an avid Singapore-based hybridiser during the 1920s and 1930s. He is best remembered for *Arachnis* Maggie Oei (registered in 1940), which laid the foundation for the export orchid flower industry, *V.* Miss Joaquim being too fragile to be exported. The red Spider Orchids from the Singapore Botanic Gardens (*Arnth.* James Storie, *Arnth.* Anne Black and *Arnth.* Bloodshot) further bolstered the cut-flower trade of that period.

When a tetraploid parent is bred to a diploid parent, the resultant hybrid is triploid (for instance, all the *Vanda* JVB hybrids, and *Vanda* Nellie Morley). The hybrid is extremely floriferous but is sterile. It serves the purpose well if one's intention is to produce a cut flower of a very free-flowering hybrid for the gardener. The approach has been extensively exploited for *Aranda*. Breeders who want their hybrids to be stepping stones along the path of progress, rather than dead ends in themselves, avoid using tetraploids. Roy Fukumura said that he shies away from tetraploids for this very reason. Consequently, Fukumura's hybrid, *Ascocenda* Yip Sum Wah, in addition to earning an extraordinarily high number of awards, including a First Class Certificate, also parented a whole range of great hybrids. His *Ascda.* Peggy Foo is also extremely productive of fine hybrids. These two *Ascocenda* are examples of the perfect hybrid.

But Fukumura reminded us that despite endless endeavour and research, nothing is guaranteed when one makes a hybrid. Every once in a while, even an experienced breeder 'makes papaya' (a bad orchid hybrid that is conceptually flawed). The gamble makes raising an orchid seedling all that more exciting.

There is one last area that has been largely unexplored: breeding scent into orchids. An ancient Indian simile says that a woman without virtue is like a flower that has no fragrance. Remember that fragrance was the reason for growing *Cymbidium* in China 2,500 years ago, and the scent of *Neofinetia* was still the rage when the Shogun ruled Japan during the Edo Period. Hence, one should not overlook scent in the selection of orchids to grow.

Below: *Ascda.* Gold Delon x *Ascda.* Peggy Foo

Among the popular orchid species, *Aer. lawrenceae*, *Aer. odorata*, *Arach. flos-aeries*, *V. dearei*, *V. denisonniana*, *V. lamellata*, *V. luzonica*, *V. tessellata*, *V. tricolor*, *Phal. bellina*, *Phal. violacea*, various species of *Cattleya*, *Den. anosmum*, *Den. chrysotoxum*, *Den. crumenatum*, *Den. leonis* and, of course, *Vanilla* are strongly scented. Their fragrance is most evident in the morning when the humidity is high.

This fragrance breeds through into some of their hybrids. In the case of *V. lamellata*, *V.* Mimi Palmer HCC/OSSEA and *V.* Overseas Union Bank are scented; with *Vanda denisonniana*, the combination with *Aerides lawrenceae* boosted the scent.

This is certainly an area well worth exploring. We could perhaps see better results in the field if the molecular biologists show some interest here.

Right: *Rhynchorides* Memoria Suranaree (*Aerides lawrenceae* x *Rhyncostylis coelestis*)

Opposite: Another Thai hybrid that is strongly scented — *V. denisoniana* x *Aer. lawrenceae*

Chapter 10

Colorama and Mutations

Colorama was the name bestowed on the first gaudily coloured, splash-petal *Laeliocattleya* hybrid by Mrs. Introini of California in 1962. She chose this name for the unique cross made by Franklin W. Gamble because "that is what it looked like". *Lc.* Colorama is basically a purple-lavender *Cattleya* with a large yellow splash on its lip, but its uniqueness lies in the fact that the yellow splash is also carried on the petals (see p. 303).

The modern hybrid that most resembles this plant is *Lc.* Taiwan Queen.

Frank Fordyce referred to them as the "colorful clowns of the orchid world". He wrote an article on such hybrids in 1980 in which he paid tribute to Kay Francis, who took up the challenge to improve on *Lc.* Colorama.

How *Lc.* Colorama came about is a mystery because Gamble did not disclose how he came upon the novel hybrid. In his article on the colourful Coloramas in the August 1985 issue of the *American Orchid Society Bulletin*, Ernest Hetherington, the famous hybridiser of *Cattleya*, made some thought-provoking observations. This chapter has drawn extensively from his article.

When Gamble was dying from cancer, he offered his notes to Mrs. Introini on the undertaking that she would not pass on the information to others. She declined as she could not accept the restriction. The notes were destroyed. Hence, the theories surrounding the making of *Lc.* Colorama are all speculative.

In 1956, Gamble registered *Lc.* Peggy Huffman, a flared-petal, pink and purple hybrid that was a cross between *Lc.* Princess Margaret and *C. intermedia*. Hetherington postulates that the *C. intermedia* was surely the variety *aquinii*, possibly a diploid clone. This splendid splash-petal variety was discovered in 1891, and existing plants are derived from sibling crosses among the three original clones. (Some clones are now properly designated, but it was not the case in the 1960s.) The *Lc.* Princess Margaret was probably a tetraploid. *Lc.* Colorama is *C.* Arctic Snow x *Lc.* Peggy Huffman. *Lc.* Arctic Snow is an unexceptional pure white albino.

So how did Gamble manage to produce the splendid hybrid? Gamble knew that colchicine could increase ploidy in orchids, and there is speculation, but no absolute proof, that he might have experimented with the chemical to produce the Colorama seedlings. If he did, he must have been the first to succeed. In her correspondence with Hetherington, Mrs. Introini said that the Gamble seedlings of *Lc.* Colorama came into bloom one after another, and they were about 80 percent consistent in shape and colour — just what one would expect for a community of triploids. "My (Mrs. Introini's) first bloom came from a community pot that Mr. Gamble gave me ... It had distorted petals. The next six were beautiful, but the last three were like the first one with distorted petals." This is what one would expect from a mixed community of triploids and aneuploids. These observations suggest that one of the parent plants, almost certainly *Lc.* Peggy Huffman, had high ploidy. It might have been a fertile triploid or pentaploid, or a tetraploid. It has

Left: *Lc.* Taiwan Queen

since produced other splendid hybrids (e.g. *Blc.* Lyonors and *Lc.* Galaxy.)

Past experience had shown that when *C. intermedia* var. *aquinii* was used as a parent, it could transfer the colour and pattern of the labellum and throat to the petals in an interesting manner, but other effects were not pleasing. Sometimes the sepals became ruffled. Often, the petals swept forwards, closing over the column, giving the appearance that the flower was not open. This may affect the blooms of some plants periodically, but not all of the time. Hence, all these features manifested in *Lc.* Colorama, but excellent clones could be selected, and they continued to breed true.

Further Hybidisation with *Laeliocattleya* Colorama

Lc. Colorama 'The Clown' was a vigorous plant that was also a productive parent. Kay Francis (Mrs. Fred) of Pasadena, California, bred the first of its progeny, *Lc.* Prism Palette (x *C.* Horace 'Maxima' AM/AOS), and registered it in 1973. Here, clear white sepals and a broad white border on the petals accentuated the colourful central streak of yellow that merged into purple on the petals and similar splashes on the lip. In some clones, the base colour was a faint lilac instead of white. It was a new flower type and an instant hit, and it garnered numerous awards worldwide.

Kay Francis' generosity with her *Lc.* Colorama 'The Clown' soon permitted several hybridisers to breed it to the different colour forms of *Cattleya*. *Blc.* Frank Fordyce tested the use of a tetraploid yellow *Blc.* Golden Slippers. The surprising result was a deep purple flower with an intensified yellow splash on the petals. Another breeding that used *Lc.* Pirate King 'Port Wine' (whose ancestry is half red and half yellow-bronze), a very dark red-purple flower produced a very dark

red purple *Lc.* Petticoats with brilliant yellow flaring in the lips and petals. *Lc.* Color Guard resulted from a crossing of *Lc.* Colorama var. Caprice to a tetraploid purple *Lc.* Drumbeat. The flowers were well shaped and of a light purple hue with a large yellow splash on the petals. A similarly coloured flower is *Blc.* Silk Slippers (*Blc.* Pamela Hetherington 'Coronation' x *Lc.* Colorama 'The Clown').

Kay Francis also bred *Lc.* Colorama back to *C. intermedia* var. *aquinii* and obtained a better-shaped flowering with more intense but less interesting colours.

What do we know about other crossings with *C. intermedia* var. *aquinii*?

When its pollen was used on *Slc.* Precious Stones, an orange-red mini *Cattleya*, two-thirds of the progeny bore peloric flowers and a third bore normal flowers with grape purple petals and sepals. The sepals of the peloric flowers were a rust brown covered with purple spots, and the lips and petals are a very dark purple splashed with canary yellow. The *Slc.* Precious Stone used in this breeding was the var. 'True Beauty' AM/AOS, a tetraploid discovered among the mericlones of the original *Slc.* Precious Stone. It was a relatively simple hybrid made up of only three species, each used only once. Colorama breeding has resulted in other remarkable

Left, Top: This complex hybrid, *Hawkinsara* Sogo Doll (*Laeliocatonia* Peggy San x *Slc.* Katsy Noda), expresses the splash-petal influence of *C. intermedia* var. *aquinii*.

Left, Bottom and Opposite: There is just a hint of the splash pattern in this *Laeliocattleya* hybrid (left, bottom), but it shows the other (undesirable) effect of *C. intermedia* var. *aquinii* — i.e. the tendency for the under-extended petals to point forwards. However, that is not always unattractive, as demonstrated by this hybrid on the facing page.

Colorama and Mutations 301

hybrids in the mini *Cattleya* range. Here, mention may be made of *C.* Interglossa, a producer of many splash-petal hybrids, *Lc.* Hauserman's Firewings, *Blc.* Pride of Salem 'Talisman Cove' (which are the immediate progeny of *C. intermedia* var. *aquinii*), *Slc.* Fire Fantasy 'Hihimanu' (*Lc.* Hawaiian Fantasy x *Sophronitis coccinea*), *Lc.* Angel Heart and Lc. Mari's Song (*Lc.* Irene Finney x *C.* Cherry Chip). With the large, red *Slc.* Anzac, it produced a world-class raspberry petal *Slc.* Empress of Mercury var. Gwo-Luen AM/AOS. Crossing to the yellow *Blc.* Buttercup resulted in a yellow, red-fringed *Blc.* Horizon Flight; to the yellow, red-lip *Blc.* Ruth Witbeck, a bicoloured yellow with red veins on the petals and wide red border; to the yellow, red-lip *Lc.* Amber Glow, a mauve

coloured *Cattleya* with a central yellow splash on the petals (*Lc.* Solitude). The crossing of the white *C.* Bob Betts produced *C.* Margaret Degenhardt, a beautiful lavender Colorama-type hybrid of excellent shape. *C.* Suavior, the cross made with *C. mendelii*, is the forerunner of numerous exceptional second-, third- and fourth-generation splash-petal *Cattleyas* (*Lc.* Bella Simpson, *Lc.* Excellency, *Lc.* Red Empress, *Lc.* Uncle Sheu Wen, *Lc.* Hypellency, *Lc.* Judy Small and *Lc.* Sedlescombe). A throwback to a chartreuse ancestor (*Blc.* Xanthette) produced the unusual green, red splashed *Brassolaeliocattleya* (*C.* Moscombe x *Blc.* Golden Slippers) with parents that were phenotypically yellow. Taiwanese breeders have carried the line further (*Lc.* Mem. Dr. Peng, *Lc.* Chun Yueh, *Lc.* Moscombe,

Lc. Hong Sie Chen, *Lc.* Chiou-Jye Chen, *Lc.* Prem, *Blc.* Chinese Beauty, *Slc.* Raincombe, *Slc.* Yeong Huei Chen, etc.). Colorama hybrids now involve other genera within the *Cattleya* alliance, such as *Laeliocatonia* (*Lc.* French's Cheek-La x *Ctna.* Jamaica Red "Devine' HCC/AOS) and *Ltna.* Ernest Davidson 'Talisman Cove' (*Ctna.* Quest's Millenium x *Lc.* Aloha Case), *Otaara* Hwa Yuan Bay 'She Shu' AM/OSROC (*Lctna.* Peggy San x *Blc.* Sunset Bay) and *Hawkinsara* Sogo Doll.

C. intermedia var. *aquinii* is not the sole source of all splash-petal *Cattleya*. The beautiful splash-petal *C.* Penny Kuroda, *Lc.* Hawaiian Flare and *Lc.* Hawaiian Variable have 6–9 species in their background, but apparently no *C. intermedia* var. *aquinii*.

Mutations

Mutations following mericloning have produced a variety of interesting colour forms in *Dendrobium*, *Cattleya* and its relatives, *Doritis*, *Doritaenopsis* and *Phalaenopsis*. First noticed during the 1970s, two early examples were featured in 1982 in *A Joy Forever*, the book on *Vanda* Miss Joaquim, Singapore's national flower (Times Editions, revised, 1998). Mutation does not affect colour alone: sometimes it affects the form of the flower and even the shape and coloration of the leaves. The splash petal (apple-green splashed with red), red-lip *Lc.* Cuiseag (*C. luteola* x *Lc.* Ann Follis) is a mutation which arose from meristem culture. Among the beautiful examples of mutations within *Phalaenopsis*, mention must be made of *Phal.* Golden Tris 'SYK Magic', *Phal.* Taisuco Yellow Boy, *Phal.* Sogo Charm, *Phal.* Little Mary, *Phal.* Sogo Maryland, *Phal.* Ever Spring Light 'Three Lip' and *Phal.* Sogo Fairyland. But some have very few flowers.

Pelorism

Pelorism is the term used to describe the condition when the petals of an orchid flower share features in common with the lip. The condition is not new, but it has become more commonplace. Peloric forms appeared in a few orchids without the intervention of mericloning. *Phalaenopsis intermedia* var. 'Star of Leyte' is a case in point. Awarded by the American Orchid Society, it was written about by P.K. Manuel and T.C. Lee in the *American Orchid Society Bulletin* in 1974. *Doritis pulcherrima* collected from the wild show peloric forms with the normal (diploid), *buysonniana* (tetraplopid) and *alba* varieties, and within each variety there are several peloric types.

On these two pages, from Left to Right: *Laelia intermedia* var. *aquinii*; *Lc.* Interglossa; *Lc.* Mari's Song; *Lc.* Colorama.

Above: *C.* Moscombe x *Blc.* Golden Slippers

The peloric condition where the petals take on the appearance of a lip is called irregular pelorism.

Regular pelorism, where the lip of the orchid flower reverts to its original form and resembles the petals, is far less common. The commonest example is *Dendrobium* D'Bush Pansy, which exists as a peloric form in several tints of mauve. Some pansy-*Dendrobium* hybrids retain their pelorism, but not all. In 1997, T.D. Amore and Kamemoto explained that this was due to a single recessive gene mechanism. Thus, a back-cross to one parent is required for the pelorism to manifest.

Irregular splash petals are the second group of mutation to appear in mericlone orchids. In 1980, Koh Keng Hoe found two *Dendrobium phalaenopsis* hybrids showing such petals. In one, the dark magenta sepals and petals were sharply outlined by a rim of white. In the second clone, dark magenta flared out from the centre over a basal layer of lighter magenta. The streaking also affected the lip. In both hybrids, the demarcation between the two colours was sharp, and the outline was irregularly saw-toothed. However, the size and shape of the flowers, and the vigour of the plants were inferior to their siblings.

Streak patterns have now been encountered in other *phalaenopsis*-type *Dendrobium*, such as *Den.* Irene Chong, *Den.* Jenny Ang, *Den.* Genting Blue, *Den.* Panda, *Den.* Ekapol and *Den.* Peacock, and also in intermediate *Dendrobium* like *Den.* Kasem White, *Den.* Chaisri Flare, *Den.* (Anucha Flare x Chaisri Gold), *Den.* (*compactum* x Burana Fancy) and *Den.* Jackie.

A variation was found in *Den.* (Mary Trowse x Tomie Drake). Here, the dark pink petals edged in red resemble the frilly red lip. This is less obvious in the lighter varieties where the symmetrical sepals are of a very much lighter tint, sometimes almost white. The lemon-green stripes bordering the white petals echo the coloration of the sidelobes of

the lip in *Den*. Kasem White, a mericloned *Dendrobium*.

In addition to streaking, the flowers of *Dtps*. Ever Spring Prince, which open at different times, show dissimilar colour patterns on the same inflorescence. This chimeric phenomenon is commonly encountered in *Phal*. Golden Peoker and its hybrids. It is due to somatic mutation, engineered by status of the plant, and may possibly be related to a chromosomal or genetic instability in the hybrid. In nature, unisex eggs of the turtle develop into male or female creatures depending on the ambient water temperature; that is somatic mutation.

Fasciation

Mutation sometimes manifests as fasciation, the condition where multiple petals and/or lips are crowded together on a flower. The term is derived from the Latin *fasciculum*, meaning

Opposite, Top: The small red streak at the tips of the petals in this *Brassolaeliocattleya* hybrid (*Blc*. Tainan Gold) is not quite a splash-petal effect. It is unrelated to *C. intermedia* var. *aquinii*

Opposite, Bottom: *Cattleytonia* Varut Chrystal (left); *Lc*. Phra Nakhorn Khiri (right)

Above: This recent Thai hybrid attempts to combine heavy substance, floriferousness and splash-petal effect.

Top, from Left to Right: *Doritis pulcherrima*, three peloric forms

Right: *Phal. intermedia var.* 'Star of Lyte'

Opposite, Top: *Dtps.* Ever Spring Prince (mutation)

Opposite, Bottom Left: *Dtps.* Little Mary (mutation)

Opposite, Bottom Right: *Dtps.* Ever Spring Prince (mutation 2)

'a bundle'. The phenomenon is not confined to orchids. The Chinese cockscomb (*Celosia* or Ten Thousand Year flower), so beloved during Chinese New Year; the cauliflower; the sacred myrobalan (*Citrus medica*, the so-called Buddha's Hand); and various cacti are examples of fasciated fruits and plants. Caused by the splitting and proliferation of apical meristems, the process can be triggered by growth hormones, chemicals, radiation, viruses and other infectious agents. Two oriental *Cymbidiums* (*Cymbidium sinensise* and *Cymbidium goeringii*) are prized for their fasciated flowers. In his discussion of the topic in *Orchids* (July 2001), the magazine of the American Orchid Society, Carl Withner

mentioned that a Belgian horticulturist crossed two peloric forms of *Cattleya labiata* in the 1920s. After two or three generations, he obtained *Cattleya* flowers with 15–20 petals but without a lip. Fasciated flowers sometimes appear within an inflorescence of peloric *Phalaenopsis*. Such mutations are rarely

attractive and have not been traditionally admired in the West. However, recently, the American Orchid Society gave a Judges' Commendation Award to recognise that this trait that might be worth exploring.

Top: *Den*. Emma White x Udomsri

Right: *Den*. Pompadour

Far Right: *Den*. Ekapol (mutation)

Opposite: *Den*. Tanchai Gold

Overleaf, from Left to Right: Three interesting peloric forms arising from the mericloning of *Dendrobium* hybrids — *Den*. Kasem White; *Den*. (*compactum* x Burana Fancy); *Den*. Peacock

Peloric *Dendrobium* Hybrids

Clockwise from Top: *Den.* Thong Chai Gold (mutation); *Den.* Ekapol; *Den.* Yuan Nan Dreamy; A new Phalaenopsis-*Dendrobium* mutation

The *Den.* Yuan Nan Dream, a *Den.* D'Bush Pansy-type, is the sole example of regular pelorism, and it has produced hybrids that also show similar pelorism. The other *Dendrobium* mutations are referred to as irregular pelorism.

Opposite, Clockwise from Top Left: *Den.* Anucha Flare x Chaisri Gold; *Den.* Panda; *Den.* (*canaliculatum* x *biggibum*) x *Den.* Liholisan; *Den.* Chaisri Flare.

Chapter 11
Standards of Quality

Visual impact draws one's attention to a stunning face in a crowd, to an exceptional work of art in an exhibition, to a single flower in a bouquet or to an orchid spray in a collection. In an orchid flower, visual impact is the result of the interplay of the following characteristics: brightness, intensity and clearness of colours, as well as contrast of colours with distinct colour patterns; outstanding form, including elements such as balance, roundness or fullness and flatness; novelty; unusual texture and graceful arrangement of the flowers on the stem; and exceptional size, substance and floriferousness. Quality is the combination of all these desirable attributes in a single orchid.

The standards of quality erected by orchid societies throughout the world are based on the above criteria, with varying weightage being placed on each criterion, depending on the taste of the centre in question and the genus involved. The oldest awarding body is the Orchid Committee of the Royal Horticultural Society, which has ranked its awards according to merit with First Class Certificate (FCC), Award of Merit (AM) and Highly Commended Certificate (HCC). This terminology has been adopted by all awarding committees. There is a considerable degree of consensus regarding the ideal flower, although the standards of excellence which are regarded as being worthy of an award do vary from one centre to another. The plant's rarity always acts in favour of the plant, while very high standards are expected for orchids that are commonly cultivated.

When considering a plant for a quality award, the judges consider a single spray; of a plant, the best spray is judged. The other sprays need not match up to the same standard. On a different occasion, the flowers may improve, in which case the plant may be resubmitted. More commonly, particularly if the plant is no longer receiving the same quality of care, the flowers may not reach the standard of the spray that earned the award.

The oldest awarding organisation in Asia is the Orchid Society of Southeast Asia (OSSEA, formerly known as the Malayan Orchid Society), now in its 77th year. It is interesting to note that it took almost 40 years for the OSSEA judges to award their First Class Certificate, and certainly this was not due to the lack of fine plants, for several had won the FCC from awarding societies abroad when their flower sprays were sent there. And so far, OSSEA has only made one FCC award. Awards from the following societies are highly prized: the Royal Horticultural Society, the American Orchid Society, the Orchid Society of Southeast Asia (formerly the Malayan Orchid Society), the Orchid Society of Thailand, the Hawaiian Orchid Society, the South Florida Orchid Society (their Silver Medal is the equivalent of the Award of Merit), the Australian Orchid Council, the Japan Orchid Society and the Orchid Society of the Republic of China.

Some societies have specialised expertise in particular areas. The Orchid Society of Southeast Asia, for instance, is most familiar with Spider Orchids. Some societies also set higher standards than others; and still other societies, especially large ones, may have stricter criteria for specific departments (zones). The awarded plant is exceptional in the eyes of the judge.

Opposite: *Aranda* How Yee Peng 'Ada' AM/OSSEA (1979) — the fourth clone of the hybrid to receive an AM from OSSEA, indeed a very rare honour

Right: *Mokara* Zaleha Alsagoff 'Nong' FCC/OSSEA (1997) (Photo by courtesy of Syed Yusof Alsagoff)

The Reasons for Award Judging

The purpose of award judging is to recognise superior forms of orchids, set goals for hybridisation, designate outstanding milestones and reward growers and breeders of fine plants.

Systems of Award Judging

An orchid award committee may elect to use either the Appreciation System of the Royal Horticultural Society or the Point System that was introduced by the American Orchid Society. Under the Appreciation System, a large number of experienced judges (usually a minimum of seven) is required to judge a plant. A First Class Certificate (FCC) must be supported by a positive vote of 73 percent or more, an Award of Merit (AM) by 67 percent of the votes, and a Highly Commended Certificate (HCC) by 51 percent. With the Point System, three judges may form a quorum (although a larger number is desirable), and they will individually and independently award points on a prescribed score sheet in which the desirable floral characteristics, such as colour, form, arrangement, floriferousness, size, substance, texture, rarity, etc., are itemised and apportioned their percentage of the total points. An average total score of 90 percent and above qualifies for a First Class Certificate (FCC), while 80–89.9 percent qualifies for an Award of Merit (AM) and 78–79.9 percent a Highly Commended Certificate (HCC). Some societies lower the HCC requirements to 75 points, but this generally makes the judges stricter in their assessment. When a large proportion of the judges in an award committee are too severe, the Appreciation System may fail altogether, and only the Point System will enable at least some plants to win awards each year.

In considering a plant for an award, the award judges of the Orchid Society of Southeast Asia also ask themselves this question: Is this plant an improvement on the previously awarded clones? The point system does not make this requirement, and it is up to the judge to remember that the purpose of judging is to encourage continuous improvement in the quality of cultivated orchids. Hence, the search is for better plants and not for plants which equal the standard set before. To reinforce this idea, every awarded plant in the present publication bears the date of its award. This makes it simple to compare awards of different decades and to observe progress.

Opposite: *Aranda* Hee Nui 'Khoo' AM/OSSEA (1967)

Below: *Aranda* Wong Bee Yeok 'Maryland Nursery' AM/OSSEA (1975)

Right: *Trevorara* Ian Trevor 'Nong' HCC/OSSEA (1970)

The Identity of the Award Plant

The awards are given to specific plants (not to the crossing), and these are then designated by varietal names suggested by the owner, such as *Vanda* Tan Chay Yan var. 'Pride of Malaya' FCC/RHS. All simple vegetative propagations (cuttings and offshoots) from this plant will carry the same award since they are identical to the original plant. There is still no agreement on the status of mericloned seedlings of awarded plants because some variation is often present. The proof of pedigree is the award certificate presented by the awarding society to the grower who submitted the plant for the award. A photograph and a detailed description of the awarded orchid is always published in the journals of the orchid society in question and these may be consulted to check on the identity of a clone.

Whenever the name of an orchid is accompanied by its award, it is highly recommended that the date of the award also be included. This ought to be standard practice for every orchid journal.

Award Judging by the Orchid Society of Southeast Asia (OSSEA)

The Orchid Society of Southeast Asia switched from the Appreciation System to the Point System in 1962. Out of a total of 100 points, 30 are given for colour, 30 for form and

Left: *Mokara* Mak Chin On 'Hui Lan' AM/OSSEA (1980).

Opposite: *Vanda* Tan Chay Yan var. Tan Yeow Cheng AM/MOS (1959) is still a beautiful plant by today's standards.

shape, 20 for stem characteristics and 20 for other floral characteristics, particularly size, substance and texture. Because only a single spray is judged, such desirable horticultural characteristics as the free-flowering nature of the plant, its ease of cultivation or the skill in growing and flowering a temperate plant are all not assessed. Neither, for that matter, are scent, lasting qualities of the bloom and its ability to travel as a cut flower considered. A spray must be more than 60 percent bloomed but not over-bloomed, and it must be free from blemish before it will be considered for an award.

The judges should not be aware of the ownership of the plant. A judge who is the breeder or the owner of the plant refrains from making comments on the plant and is not involved with the judging.

Desirable Qualities

Colour must be bright, clear and intense. Rarity is prized, as in the case of albino and two-toned forms. Where more than one colour is present in a significant amount, the colours must harmonise. If patterns are present on the petals and sepals, such as tessellations, spotting or striping, they should be distinct. A natural blush towards the centre of the petals and sepals is acceptable (indeed, on occasion, it enhances the beauty of the flower), and a fine distinct rim of white outlining the sepals and petals adds novelty, but a colour break which is a smudging and fading of colour towards the tips of the petals and sepals (often caused by virus) immediately disqualifies the plant.

With regards to form and shape, the key words are balance, roundness, fullness and flatness. That said, the horn-shaped *Dendrobium*, the Scorpion Orchids and the flaming *Renanthera* are admired for their peculiar shapes; balance is the important criteria here, and fullness is expressed in broader sepals

and petals. In *Vanda*, *Ascocentrum*, *Phalaenopsis*, *Dendrobium* (*Phalaenopsis*-type), *Cymbidium* and *Paphiopedilum*, one would like to see roundness, fullness and flatness. The floral segments should overlap so that there are no gaps between the sepals and petals. The tips of the sepals and petals, or the lip, as in *Oncidium papilio*, should be rounded.

The stem characteristics of importance are habit (whether erect, arching, pendulous and, if branching, well fanned out), length, stoutness, arrangement of the flowers on the stem and floriferousness. In the case of orchid species, erect, arching and pendulous inflorescences are all acceptable, but in orchid hybrids, the erect and arching inflorescence are preferred to the pendulous inflorescence. The rachis should be straight, not zigzagged or twisted. Where the rachis is branched, the side branches should be held away from the main rachis so that the flowers are well displayed and not bunched together. A tall, stout, erect stem is preferred to a short, slim and erect stem because the former will present the flowers more attractively above the growing tip of the plant. The flowers should be well arranged, either in two rows of alternating flowers facing opposite directions or in a circular whorl around the stem, facing all directions. The flowers should not be crowded together on one side and sparse on the other. The flowers in a row should be touching or just overlapping so that there are neither wide gaps between the flowers nor overcrowding to the extent that the individual shapes of the flowers cannot be easily discerned. However, it is better to have gaps than to have flowers too bunched together.

With regards to size, substance and texture, these should be better than average for the orchid under consideration.

When two species or hybrids are interbred, the result should be an improvement on the parents before it can be considered for an award. Obviously, the resultant hybrid cannot

Judging Point — Arrangement

Above: The correct arrangement of *Phalaenopsis* flowers is illustrated by this spray of *Dtps.* Kristine Teoh var. Phaik Khuan which won HCC/OSSEA in 1971.

Opposite, Top and Bottom: When this lovely white *Phalaenopsis* was in bud, the inflorescence had received more light on one side, resulting in a reasonable arrangement on one side but not on the other; the top photo shows the side facing the light. A graceful continuous arc is formed by five flowers but there is one stray downward-facing flower below the arc. This implies that the arrangement is not balanced. The bottom photo shows the obvious poor arrangement on the reverse side. The spray will not receive an award at this stage. The plant may be resubmitted when it is properly grown with respect to ambient light. The problem is commonplace in *Dendrobium* where it is not so well appreciated because the inflorescence may be held against the plant.

be either larger or more floriferous than both parents because this is not the way the Law of Geometric Averages works in nature, but the awarded plant should have more flowers and/or larger flowers than is average for that particular hybrid to earn high points in that section of the score sheet.

If a species or hybrid has already been awarded, a subsequent submission must surpass the awarded clone to qualify it for an award. (This rule may not apply in some awarding bodies.) If a plant is doing better than when it was originally awarded, it may be resubmitted for

Judge a flower on its merit, not on its initials.

Above, Left: *Phal. violacea* (*Phal. bellina*). The plant does not carry an award because it had never been submitted for judging. Nonetheless, it is undoubtedly of award quality.

Above, Right: *Phal. violacea* var. Neo Beauty HCC/OSSEA (1980). An outstanding and well grown plant. No one would quibble about granting an award to this plant.

the judges to decide if they will give it a higher award; in other words, the award recognises the particular orchid plant at its maximum potential. If the plant is not well looked after, its subsequent flowering may be poorer.

Other Awards of OSSEA

The Award of Botanical Merit (ABM) given to species is the equivalent of the Award of Merit. In addition, OSSEA also confers a Certificate of Cultural Commendation (CCC) to plants which are so well grown that they are extremely free-flowering. Here, a small, healthy plant outshines a large plant which is equally floriferous.

There are many outstanding plants which have never won an award because they were never judged. The reason could be that their owners were unaware of the award potential; or that they bloomed at the wrong time; or that they were insufficiently well grown, or that the flowers had been damaged by insect attack or inclement weather. Sometimes it was because their parentage was uncertain, and at other times, although rarely, because their owners did not care.

If you own a plant you consider worthy of an award, try to grow it well. Protect it from rain, pest and disease, and you can submit it at its third or fourth flowering after re-potting. Learn to time its flowering to coincide with a judging session. If necessary, orchid flowers can be sent by air freight to all parts of the world for presentation before orchid committees. Never submit a plant if you think that it will do better later on; it might just win you an HCC instead of the coveted AM.

If you are buying a plant, an award means that the plant's quality has been recognised by the experts. Nonetheless, check the year. If you see a top rate plant, trust your own judgement; ignore the lack of initials and grab the plant.

The growing condition has a profound effect on the quality of the flowers.

Top, Left: This *Cattleya,* which carries an FCC/AOS, flowered in Singapore. Flower size is much reduced and the petals are a bit narrow; nevertheless, the quality of the flower is evident. It does not have an OSSEA award because the plant does not reach its maximum potential in Singapore.

Top, Right: *Pot.* Buruna Beauty 'Burana' AM/OST retains its beauty when it flowers in Singapore.

Above, Left: *Lc.* Alma Kee AM/AOS, grown and flowered in full sun in the author's garden. The exposure to strong light and high temperature has compromised the shape of this lovely *Laeliocattleya* which is a reliable bloomer in Singapore.

Above, Right: *Blc.* Chunyeah var. Tzeng Wen AM/AOS was grown in Taiwan and brought in bud to Singapore for a show. The flower has retained its perfect form.

Chapter 12

Control of Flowering

For an orchid grower, the finest moment arrives when a new hybrid puts out its maiden blooms. If he has several plants of this hybrid, once the first plant blooms, he will be on his toes awaiting the flowering of the rest. The commercial cut-flower grower with several acres of orchids, unless he also happens to dabble in hybridisation, probably does not share the enthusiasm and eagerness of the amateur. For him, happiness is getting maximum quantity at the right season, so timing and yield are prime concerns. A knowledge of the factors which control the flowering of orchids can be employed by all orchid growers, whatever their interests, to:

1. Shorten the interval between seed germination and flowering;
2. Time the flowering for orchid shows, festivals and periods of peak demand;
3. Stimulate flowering in plants which are persistently vegetative; and
4. Assist in the proper selection of suitable plants which will flower freely in one's growing environment.

Ripeness to Flower

Although some orchid seedlings have flowered while still in flask, and some in compots, as a rule, the stimuli which evoke flowering in adult plants are incapable of inducing flowering in immature seedlings. There is a minimum size which each orchid plant must reach before it will respond, attaining 'puberty', as it

Left: Gregarious flowering of *Coelogyne rochussenii* grown by Mak Chin On

were, and become receptive to the flowering stimulus. The minimum time between seed sowing and flowering varies from one genus to another. Within a similar hybrid crossing, it is influenced by the choice of the parental clones; in particular, crossing it will vary from one plant to another.

Puberty is reached in sympodial orchids when the first mature pseudobulb is produced. With good culture, this is usually the third or fourth pseudobulb after placement into individual pots.

The leaf-time scale (the plastochrone scale) is employed to assess maturity in monopodial orchids. In each crossing, the seedlings need to grow a minimum number of mature leaves before they will flower: three in *Phalaenopsis*, eight in *Vanda* and 14 in *Aranda*. In hybrid orchids, the ability to flower at low height is strongly influenced by parentage. For example, the early *Aranda* produced in the 1940s did not flower until they had 30–40 leaves (which took 5–6 years) whereas modern *Aranda* blooms at around 14 leaves, taking only 2.5–3 years. Different cultivated varieties (cultivars) within a crossing differ in their rate of growth, given the same environment, and one may reach puberty, say, the 14-leaf stage, six months ahead of the rest. It will flower first but the remaining cultivars will also bloom when they reach the 14-leaf stage, although only six months later. Conversely, two plants may flower at the same time, but at different stages of development, say, at the 14-leaf and 20-leaf stages. Improved cultural techniques have been worked out to accelerate vegetative growth to achieve puberty in a shorter time. This can be done by judicious transfer within the shortest possible time into enriched media in flasks and

325

optimum growing conditions such as a high growing temperature, high humidity, adequate and continuous lighting, frequent heavy fertilising and the use of organic fertilisers. The nurserymen in Thailand are setting new standards in this area. It is now not uncommon to see large strap-leaf *Vanda* flowering on a plant that is less than 15 cm tall.

Bud Primordia

When a flowering stimulus reaches a plant, only those resting buds, or bud primordia, which are capable of developing into inflorescences, start to grow and differentiate. Orchids can be divided into two broad categories according to the position of these bud primordia.

Pseudobulbous Orchids with Single Terminal Bud Primordium

In these orchids, such as *Paphiopedilum* and *Cattleya*, each pseudobulb produces a single inflorescence and flowers only once. Some may not even flower. The equitant *Oncidium* also has a single bud primordium and each plant will produce a single inflorescence which is capable of producing successive crops of flowers from dormant buds on the rachis.

Orchids with Multiple Bud Primordia located at the Leaf Axils

These may be pseudobulbous orchids such as *Dendrobium*, *Eria*, *Cymbidium* and *Oncidium*. All monopodial orchids belong to this category. In the case of *Cymbidium*, each pseudobulb will flower only once but can produce several inflorescences at one time. The pseudobulbs of the evergreen *Dendrobium* may flower several times a year. In *Dendrobium phalaenopsis*, a young pseudobulb flowering for the first time produces a 'terminal' inflorescence at the axil of the uppermost leaf. When the plant flowers again, this pseudobulb (now a back bulb) will produce several subterminal inflorescences which appear together with the terminal inflorescence on the new, young pseudobulbs. This flowering trait is seen in many of its hybrids. The cool-growing *Dendrobium* flowers simultaneously at every bud primordium on the distal half to two-thirds of old, mature pseudobulbs after a cool, dry, resting stage.

The monopodial orchids generally flower at several leaf axils at a set distance below the growing point. This may be four leaves in the case of *Arach.* Maggie Oei; at the other extreme the plant may only flower from the basal leaf axils as in the *Paraphalaenopsis* and *Phalaenopsis*. As the plant matures, increasing numbers of inflorescences are produced at each flowering. A curious characteristic of the *Eu-Phalaenopsis* is that two bud primordia are present at every leaf axil, one above the other. When a flowering stimulus is received, only the upper bud primordium develops into an inflorescence. The lower bud primordium does not develop unless the growing apex of the plant is removed or injured. It will then develop into a vegetative shoot. In other vandaceous species, unflowered bud primordia separate each batch of inflorescences and these unflowered buds develop either into offshoots or into inflorescences when the apex is removed.

Factors Controlling Flower Bud Initiation

The first experiments on the control of flowering in plants were conducted as early as 1920, but it was not until 1943 that E. Johnson began the first studies on orchids. Our present understanding of flower induction in orchids by controlling day length (photoperiodism) and by lowering the temperature (vernalisation) are derived primarily from the work conducted in 1951 by Gavino Rotor Jr.

Photoperiodism

Many important agricultural crops and ornamental shrubs (including orchids) which are from the equatorial region are extremely sensitive to variations in day length and respond to short days by initiating flower buds. This is true of all *Cattleya* and the *Eu-Phalaenopsis*. They are known as short day plants (SDP). In the northern hemisphere, the *Phalaenopsis* are stimulated by the short days of November and flower in February. *Cattleya triaeniae* initiates its flower buds in mid-September but waits until the even shorter days of November before the buds start growing in the sheath. *Cattleya mossiae* initiates its flower buds at the beginning of November but requires two months of short days and cool nights (around 13 degrees Centigrade) to start them growing in the sheath. *Cattleya gigas* initiates its buds during the short days of February, but only if the night temperature is below 13 degrees Centigrade; if the temperature is higher, it will not flower. However, once the buds have started growing, neither a rise in temperature nor an alteration of day length will prevent flowering.

Cattleya labiata initiates flower buds during the longest days of midsummer, but the buds remain dormant and only resume growth in the fall, when the days are short and the nights are cool. Under experimental conditions, *Cattleya labiata* will not flower if the day length is long and the night temperature exceeds 18 degrees Centigrade. It will flower if either the day length is shortened or the night temperature is dropped to below 13 degrees Centigrade, or if both these conditions are present. Thus, it is also a short day plant.

Opposite: *Grammatophyllum speciosum*. Its flowering is unpredictable but spectacular.

Some plants which are native to the temperate region will only flower if they have been subjected to long days, and they are referred to as long day plants (LDP). These plants normally flower in autumn but they will flower seasonally if they are given supplementary lighting to extend the day length.

Most hybrid orchids which grow in the tropical lowlands appear to be uninfluenced by day length, and thus, they are probably day neutral plants (DNP). The critical factor for good flowering of such orchids is high intensity light. Shading invariably reduces their flowering potential. In experiments at the University of Hawaii, Dr. Sheehan, Dr. Murashige and Dr. Kamemoto found that the flowering pattern of two *Dendrobium* hybrids were unaffected by day length: *Den.* Lady Hay flowers from October to January while *Den.* Jaquelyn Thomas flowers throughout the year with a peak from July to November; short days, neutral days and long days do not shift the flowering season to any extent. However, when supplementary lighting (between 10 p.m. and 2 a.m.) was given to produce long days, flower production in *Den.* Jaquelyn Thomas was increased by more than 50 percent compared to neutral days.

Because we have a fixed 24-hour day, short days mean long nights and conversely long days mean short nights. In fact, short day plants really require long nights to bloom, and if they are subjected to a period of illumination during the night, flowering is prevented. In some instances, interruption of the dark period by a burst of bright light for only one minute effectively prevents flowering.

With respect to the light period, it was found that the SDPs require a high light intensity even during short days to flower properly, and that their floriferousness is, up to a point, directly related to the amount of light they receive. This is a reflection of the need for a continuous, active supply of energy through photosynthesis because (1) carbon dioxide is necessary during photo-period if flowering is to occur, and (2) spraying sugar on the leaves enables SDPs to bloom in complete darkness. In *Cymbidium*, it was observed that the more light received during the growing period, the greater the flowering response to cold. Thus, *Cymbidium* are taken outdoors from late spring to early fall to enable them to receive more light, and they bloom better than those plants which are grown continuously in a shaded greenhouse in the same locality. Singapore orchid growers have known for a long time that the more light their orchids receive and the higher the temperature, the better the flower production.

Vernalisation

For many orchids, temperature variation has a decidedly more pronounced effect than day length on flowering. *Dendrobium crumenatum*, which flowers gregariously, is known to respond to the sudden brief fall in temperature associated with thunderstorms. All the plants in one locality will bloom nine days after the storm, the flowers lasting only from dawn till mid-afternoon. *Bulbophyllum vaginatum* also flowers gregariously several times a year and probably responds to a similar temperature stimulus. *Bromheadia finlaysoniana* responses likewise to a drop in temperature.

'Chilling' is necessary for the flowering of many orchids whose natural habitats are more than 500 m above sea level. Mature *Eu-Phalaenopsis* and their hybrids initiate flower spikes when the night temperature

Opposite: *Dendrobium crumenatum*

Right: *Bulbophyllum vaginatum*

is lowered to 13 degrees Centigrade for two weeks, and *Paphiopedilum*, which remain vegetative in the tropical lowlands, burst into spectacular bloom when brought into the cool house with a growing temperature of 20 degrees Centigrade. The nobile *Dendrobium* require a cool, dry season during maturation of the pseudobulbs in order to flower properly, but once they have been vernalised, they will bloom even when they are brought to Singapore. However, after the initial or first flowering, these plants become vegetative or flower very poorly. *Ascocentrum curvifolium, Ascocentrum ampullaceum, Vanda coerulea, Rhynchostylis gigantea, Vandopsis parishii, Phalaenopsis lindenii* and *Phalaenopsis schilleriana* are examples of the beautiful monopodial orchids which unfortunately require cool night temperatures to flower.

The gap between the growing temperature and the flowering temperature is probably of some importance, although no proper study has been done. In California, a night temperature of 13 degrees Centigrade is employed to stimulate flowering in *Phalaenopsis,* while in Florida, a night temperature of 18 degrees Centigrade, is adequate. In Singapore, the night temperature does not fall below 20 degrees Centigrade and even during the cool season, night temperatures are around 21 degrees. *Phalaenopsis* flowers fairly well in Singapore, although admittedly not to the same degree as in either Florida, California or Taiwan.

Interaction of Day Length and Temperature

Some orchids require both cool nights and short days to flower. Thus, most *Cattleya* species and the cool-growing *Paphiopedilum* will not flower at night temperatures above 20 degrees Centigrade, regardless of the day length.

In *Cymbidium*, low temperature induces flowering. If the day temperature suddenly rises above 18 degrees Centigrade, bud drop occurs. In *Rhynchostylis gigantea*, cool nights initiate flowering, and a cool temperature is required to achieve the proper colour intensity in the fine solid red forms known as the 'Sagarik strain'. If the day temperature is high, colour breaks appear at the base of the sepals and petals. On the other hand, the flowers of a lightly spotted common-type *Rhynchostylis gigantea* may be mistaken for an *alba* form if the plant flowers during an unusually hot season. In *Phalaenopsis schilleriana*, high night temperatures will cause the flower spike to elongate indefinitely without ever producing flowers and long days encourage plantlet formation on the flower spikes. Other undesirable side effects of high temperature on the flowers are small size, poor shape and diminished colour intensity.

Hormones

It has been amply demonstrated in herbaceous plants that the flowering principle is produced in the leaves (rarely in the stem), which are sensitive to variations in day length and temperature. The name florigen was coined to designate the flowering hormone. Subsequently, plant physiologists discovered that diverse hormones appeared to be capable of triggering the flowering response in different species, and the evidence suggests that florigen may not be a single substance.

Auxins which are responsible for vegetative growth in apical meristems almost invariably suppress flowering, but in the exceptional case of the pineapple, they stimulate flowering, a discovery which has been widely exploited in the commercial growth of pineapple. In orchids, the application of auxins does not produce flowering; instead, the removal of the natural source of auxin by the decapitation

of the growing point produces flowering in some monopodial orchids. In *Vanda* Miss Joaquim, high flower yield was directly related to the amount of sunlight received by the plants, and it was inversely related to the level of auxins in the tissues. A substance called 2, 3, 5-triiodobenzoic acid (TIBA) interferes with the polar transport of auxin, and if it is applied to a plant, it decreases the endogenous auxin concentration below the point of application. At the University of Singapore, Dr. Goh Chong Jin found that when *Aranda* Deborah was pretreated with TIBA, its subsequent response to the appropriate hormonal stimulus (benzyladenine) was greatly enhanced. Naturally occurring growth inhibitors such as abscisic acid and synthetic growth retardants such as cycocel (CCC) and B9 promote flowering in strawberries and Japanese Morning Glory, while synthetic growth retardants promote flowering in apple and pear trees. The application of these substances to *Aranda* Deborah also initiated flowering (although not to the same extent as the cytokinins), but two-thirds of the buds aborted. At the concentration required to cause flowering, these growth retardants have a severe suppressive effect on apical growth.

In long day plants (LDP) and long day-short day plants (LDP-SDP), gibberellic acid (GA3) induces flowering even when the plants are grown under short days. Several garden crops which require vernalisation to flower will bloom upon application of gibberellic acid, but this treatment has not been reported for cool-growing orchids. In Singapore, gibberellic acid was found to have no independent stimulatory effect on dormant buds, but it had a synergistic effect with 6-benzyladenine on the flowering of two *Dendrobium* hybrids and on *Aranda* Deborah.

In 1974, my wife and I observed that when benzyladenine was applied in a cream base on the dormant buds on the lower

Right: *Dendrobium spectabile* with a glorious inflorescence at the Cool House of the Singapore Botanic Gardens. Sometimes it is necessary to have a cool house to get an orchid to bloom in the tropical lowlands. This is a rear view of the inflorescence which was growing towards the light, and thus its proper face was beyond the reach of the photographer. Nevertheless, it was quite a spectacle.

Overleaf: *Vanda* Miss Joaquim at the Singapore Botanic Gardens. Henry Ridley, the Director of the Singapore Botanic Gardens who registered the hybrid in 1893 reported that it was bred by Miss Agnes Joaquim.

nodes of the inflorescences of *Phalaenopsis,* it initiated growth, and we advocated its use as a simple method for obtaining bonus plants on old inflorescences. However, we found that if benzyladenine was applied too soon after the initial flowering, the dormant nodal buds developed into side sprays of flowers. Dr. Goh also found that *Aranda* Deborah, *Dendrobium* Lady Hoochoy and *Dendrobium* (Buddy Shepler x Peggy Foo) flowered when they were treated with 10–3 benzyladenine. The hormone was injected directly into the stems of the *Aranda* and into the mature pseudobulbs of the *Dendrobium.* The treated plants flowered despite the unfavourable light conditions, which suppressed flowering in control plants. Dr. Goh also observed that the stimulatory effect of benzyladenine was accelerated by gibberellic acid and suppressed by auxins (indole-acetic acid). Kinetin had no stimulatory effect.

Flower bud development is suppressed by high day temperatures. Benzyladenine at 400 parts per million in a spray will substitute for low temperature to support bud development in these *Dendrobium.*

Nevertheless, we do not recommend the use of hormones to initiate or to maintain flowering because it sometimes results in the formation of abnormal flowers and stresses the plant, and there is always the possibility of causing permanent damage to the plant.

Stress Response

Many orchid species whose natural habitats expose them to periods of drought require to be 'dried off' after their growing season before they will bloom. Some monopodial orchids have to attain a tall height of several metres before they will flower, but they can be made to bloom at times by top-cutting (the top cut will flower if left unwatered for some time) and by decapitation (the bottom segment flowering). It is a one-off method which is only suitable for large plants that are due for replanting. Nevertheless, it is useful to meet demand for a special occasion. A deep but subtotal incision across the stem serves the same function, but several Singapore scientists found that this process resulted in plantlets as often as they produced inflorescences when they employed the method, and there was no apparent explanation for the differences in their results. (Division without removing the top portion for replanting was employed by Mak Chin On for the rapid propagation of *Aranda* Christine No. 27 when this plant was a hot favourite in the 1970s.)

Right: The Mandai Orchid Gardens is a permanent showcase of sun-loving orchids that are easy to cultivate and flower. Photographed here in the mid-1980s by the author's wife, Phaik Khuan, the author is looking for the best spray of *Vanda* Chia Kay Heng to photograph. The Garden has continued to maintain its high standards.

Flowering in Apparently Neutral Species

In the case of the ever-blooming *Arachnis*, *Vanda* and *Aranda*, new inflorescences are initiated as soon as all the blooms have dropped off the old inflorescences. However, the interval between successive crops of flowers can be reduced by simply removing the inflorescences when half the flowers on the stalks are open, a common horticultural practice.

John Ede made a detailed observation on the flowering of *Arach.* Maggie Oei var. 'Red Ribbon' in the Mandai Nursery of Singapore Orchids Ltd. between 1962 and 1963. He found that flower stalks had an average growth period of 70 days regardless of the actual length of the inflorescence, the number of flowers on the stalk and existing weather conditions. Fourteen days after he removed all the spikes from the plants, the tips of new stalks burst through the leaf sheaths enveloping the stem. The uppermost bud always appeared opposite the fourth matured leaf from the apex. However, even if only a single inflorescence were left on the stem, a new crop of inflorescences did not develop on that particular plant. Thus, if all the inflorescences on a single plant matured and were removed at 71 days, and there is a rest period of 14 days, the full cycle would take 85 days.

It is possible to get four crops of flowers in slightly over 11 months (340 days). On average, the plants produce 11.5 stalks per plant per year, two stalks at the first flowering at a height of 1.6 m, increasing to three or four stalks by the fourth flowering. There are about four unflowered axillary buds between successive crops of flowers.

At that time in Singapore, a 1 m top-cutting of *Arach.* Maggie Oei fetched a wholesale price of 80 cents to a dollar and a spray of flowers fetched 8 to 10 cents: the plant paid for itself within a year, at which time it can be top-cut. The backshoot produced 4–12 plants when it was properly cultivated. This is the type of free-flowering capacity one must employ as a

yardstick in deciding on the suitability of new hybrids for the cut-flower trade.

Several hybrids measure up to this standard — *Vanda* Miss Joaquim, practically all the quarter-terete *Vanda* Josephine van Brero hybrids, *Aranda* Christine No. 1, *Oncidium* Golden Shower, *Aranda* Baby Teoh, *Mokara* Mak Chin On and many intermediate *Dendrobium*. The modern Scorpion Orchids can be expected to produce 12–15 sprays per plant per year. The flowering cycle in *Aranda* takes only 8–10 weeks, and 5–6 crops may be obtained annually. Flowering is intensified in certain months of the year, usually March–April and September–November. Each plant then produces one or two more spikes than

is its custom. Because continuous flower production depends on photosynthesis going into top gear, it is likely that the amount of light and energy received by the plant is the decisive factor causing the peaks of flowering. Thus, long days with clear skies encourage flower production.

Nutrition

Finally, a point on the use of fertilisers. From the foregoing discussion, it seems that there is some antagonism between vegetative growth and flowering. Most orchid growers know that pure nitrogen fertilisers and high nitrogen fertilisers produce lush green foliage and soft growth but no flowers. The condition is further aggravated by inadequate lighting. A high proportion of phosphorus and, particularly, potassium stimulates flowering. For plants which are grown in direct sunlight, the use of organic fertilisers produces both vigorous growth and profuse flowering. Cattle manure produces continuous blooming in *Vanda* Miss Joaquim grown in beds on the ground, while chicken manure was widely used for *Aranda* in Penang, and pig manure was standard fare with the orchid farmers of Singapore. 'Forced feeding' for a period of 2–3 weeks after harvesting usually increases the number of spikes initiated for the next flower yield. In the case of monopodial orchids that flower continuously, the removal of immature flower spikes results in a substantial increase in flower production at the next flowering — two months in the case of *Vanda* Miss Joaquim, four months in the case of *Aranda* Christine. This practice is used by nurserymen to meet the peak demands of Mother's Day, Valentine's Day, Christmas, New Year, etc., and by some hobbyists to prepare their plants for flower shows and award judging. It is employed for warm-growing *Dendrobium* and *Oncidium*, but obviously it is not a suitable method for *Cattleya* which produce terminal buds and are slow in developing new pseudobulbs.

Numerous investigations on the flower production in commercially important orchids

that were conducted in Singapore and Malaysia have thrown up a wide range of results in the hybrids. Choy Sin Hew's tabulation of the data in Joseph Arditti's *Orchid Biology, Review and Perspectives*, VI (1994) gives the following results:

The yield (number of sprays per plant per year) for *Arach*. Maggie Oei was 15–20, 14 and 13–14, approximately a 25 percent difference between the lowest and the highest, which is quite an acceptable figure. However, for *Oncidium goldiana* (Golden Shower), four different investigators reported an annual yield of 5–6, 8–10, 10–15 and 14 respectively, i.e. a difference of greater than 150 percent between the lowest and the highest yield. This calls into question the comparisons made on the yields of the diverse hybrids studied by different investigators. No comparison can be made of the various hybrids if they are grown by different investigators, and sometimes a single investigator may not understand the needs of the individual hybrid. For instance, a slight variation in lighting may produce dramatic effects. However, we may still draw an important conclusion from the review of the studies, namely that the right growing conditions must always be found for each individual plant to maximise its yield of flowers. It is possible to do this some of the time, but it is very, very difficult to keep a plant in optimum condition all of the time.

Opposite: **Commercial white, intermediate** *Dendrobium.*

Above: **An old favourite —** *Arach.* **Maggie Oei**

This degree of floriferousness is required of a hybrid grown for the cut-flower market.

Chapter 13

Diseases of Orchids

The existence of an orchid species in abundance in any particular location testifies to its ability to adapt to the stresses of the particular environment, but it does not necessarily prove that the natural habitat has provided the best conditions for the growth of the species. This has been amply proven by glasshouse cultivation under which any orchid plant can be brought to a state of perfection, far surpassing its appearance in nature. So accustomed, indeed, are we to perfect culture that the orchidist is greatly disturbed by minor alterations in the rate of growth, infestation by parasites, damage caused by chemicals, waterlogging, dehydration and infection, all of which are quite prevalent in nature. The prevention of disease among his plants is a daily concern of every serious orchidist.

Diseases of orchids can be classified under the following categories :

1. Genetic — inbreeding, aneuploidy and polyploidy (excessive)
2. Nutritional — water (insufficient or excessive); light (insufficient or excessive); nutrients (insufficient or excessive); pH (alkalinity or excessive acidity)
3. Infections (Stress Factors) — viral, bacterial and fungal
4. Infestations — arthropods, snails and plant pests

Genetic disorders are inherent in the plants and cannot be avoided. At present, there

Opposite: There is no disease on this plant, but there is a trace of white powder at the edges of the petals due to a preventive anti-fungal spray.

is no treatment for viral disease. However, destruction by pests and fungal disease and loss of plants through poor cultural technique can easily be prevented by paying proper attention to the specific needs of the plants.

Good nursery management involves consideration of factors such as choice of site, housing, cleanliness, ventilation, humidity and rotation of fertilisers. Spraying with fungicides and insecticides also helps to prevent disease in orchid plants. The best place to grow orchids is on clear, elevated or sloping land which will provide full-day sunlight, good air circulation, excellent drainage and the widest scope for the control of light, air movement and humidity. In the tropics, orchids are usually grown either in the open or under lath houses and in whatever space one can find around the house. A new collection may at times be housed a few doors away from a neglected group of diseased plants. Visitation by disease is then continuous and unavoidable. When the orchids are in bloom, a strict insecticide programme for the control of thrips (minute insects) is essential to protect the flowers against insect damage. Other insecticides and miticides directed against the entire array of orchid pests have to be used in rotation at specific intervals. Much depends on the locality. At one time, we were growing our orchids on open land far removed from cultivated plots, and we never used pesticides or fungicides. When we moved to an old, built-up area of the city, we could barely keep up with the hordes of insects which continuously attacked our plants, and it was no longer a question of prophylactic spraying but of total war. Some friends told us that lack of biodiversity in plant forms was the cause of all our problems.

If the orchids are properly sheltered from rain, there is less likelihood of fungal disease, and again, in my experience, it is not necessary to apply frequent prophylactic spraying of fungicides. However, small seedlings which are crowded into compots, even though they may be sheltered from rain, and orchids which are grown in the open under lath houses require prophylactic spraying with fungicides once every fortnight.

Systemic and contact insecticides, miticides and fungicides can be used in combination or in rotation and should be applied through a low volume, fine mist spray to cover the plant from its growing tip to the roots in the pot or in the ground. Follow the instructions on the label. It is important to remember that these chemicals are poisonous to humans and some of the poisons may have a cumulative effect. Fungicides may produce mild to severe allergic reactions in sensitive people. Many insecticides are nerve poisons producing an immediate reaction and causing death in high doses while others may produce initially imperceptible but gradually cumulative effects with long-term usage. Singapore, the United States and many European nations have placed a ban on some of the more dangerous insecticides. I generally favour the use of only the safest insecticides for routine prophylactic spraying, such as white oil for scales, Dipterex [0,0-dimethyl (1 hydroxy-2,2,2-tricholomethyl) phosphonate; trichlorophon for short] for thrips, or liquid Derris and Nim oil which are biodegradable. The more potent insecticides should be withheld until such pests as require their specific use make their appearance. As an alternative to withholding powerful insecticides, one could use a prophylactic spraying of one or two of the following insecticides once every three months, each spraying to consist of two applications five days apart: Malathion, Dimite, Kelthane, Chlorobenzilate.

When spraying does not control an infection, it is necessary to remove the badly affected plant and trim off all those leaves, stems and roots which are affected by disease. When applying surgery to tissue which are infected with bacteria and fungi, a wide margin of healthy tissue must be removed to ensure that the cut edges are completely healthy. If necessary, even the growing apex must be sacrificed. Soon, new growth will appear at the leaf axils and one is more certain of saving a plant this way. Dead plant materials, fallen leaves and flowers and old potting medium must all be collected and burnt or be removed by refuse disposal trucks.

Genetic Disorders

Genetic disorders in plants are expressed by slow or anomalous growth, freak flowers and sterility. They may result from mutation, repeated selfing, through hybridisation, exposure to chemicals, radiation and perhaps to viruses.

The rare forms of species which are highly prized, such as the *alba* forms, are usually due to recessive traits which can only be guaranteed in the next generation by selfing the plant. Besides, selfing is also employed by growers to multiply uncommon hybrids and to improve and fix the special characteristics of the plant in question. Repeated selfing inevitably leads to an accumulation of multiple recessive traits and a loss of hybrid vigour. This makes the F5–F9 generations difficult to grow although they are predictably and consistently beautiful.

Aneuploidy refers to an aberrant number of chromosomes, and it may be the result of a crossing with a triploid or pentaploid parent. When a hybrid already possesses a mixture of chromosomes of unequal size or uneven number, it will be subfertile, and when it is successfully hybridised, there is a higher percentage of aneuploid offspring. Aneuploidy may result from mutation or chemical reaction on meristem tissue or radiation. If the aneuploid number is close to the diploid number, the plant may behave normally. If the number is far off, it will lack vigour and produce abnormal vegetative and reproductive structures.

Freak flowers and abnormal growth are not invariably due to bad genes. It is quite common for a healthy plant to produce freak flowers and abnormal leaves when it has been given an excess of fertiliser or if growth hormones and other chemicals are applied, particularly at the time of flower spike initiation. In such an event, the freaking is temporary and subsequent growth and flowering will be normal. In some instances, the anomaly is minor, perhaps affecting the arrangement of the flowers on the rachis but not the shape of the flowers themselves; if the plant has other redeeming qualities, it may continue to be grown and propagated.

Polyploidy refers to the condition where a plant possesses more than the normal two full sets of chromosomes. When three sets are present, the plant is a triploid; with four sets, a tetraploid; with five sets, a pentaploid; with six sets, a hexaploid. In animals, polyploidy is lethal or at best crippling; in plants, it appears to be an advantage, and many superior commercial crops are polyploid. Triploid orchids are extremely vigorous and free-flowering.

However, despite their abundance, the flowers are usually sterile, or if they should produce seed, a very small percentage will be viable. On the other hand, tetraploid plants are superior parents, and when they are bred with diploid plants, they confer heavy substance, size, roundness and colour to their triploid progeny. However, the tetraploids themselves are sometimes slow-growing and likely to be late-blooming seedlings.

Viral Diseases

It appears that under commercial cultivation in the tropics, nearly all orchids eventually become affected by virus. Viral diseases are spread by man, through cutting tools, and by insect vectors. There is no doubt that viral diseases eventually sap the vigour of the plant, causing poor flower production, distortion of stems, leaves, flower spikes and flowers, as well as colour break and bud drop.

In the early stages, these symptoms may be totally or partially suppressed by good cultural technique and heavy fertilising. Viral diseases cannot be eliminated through chemical sprays, and currently, there is no solution to the problem. As long as the orchids are grown by the acre in the open (and with the low prices paid for cut blooms), there is no effective and economical way to halt the transmission of viral disease. However, the orchid farmer can usually maintain his flower output by discarding old plants which no longer produce a high yield — in any case, the price for these old flowers would probably have dropped through overproduction — and replacing them with seedlings of new crosses which are virus-free. An alternative method is the production of virus-free plants by meristem tissue culture. Unfortunately, an amateur buys plants which may be infected from farmers, and the virus will soon be spread to his precious collection by insect vectors and through his use of the same cutting tools for all his plants.

The risk of transmitting virus from plant to plant can be reduced by dipping cutting tools in a saturated solution of trisodium phosphate or by flaming such tools after cutting each plant.

Badly affected plants should be discarded. An effective insecticide programme reduces the risk of virus transmission by insect vectors. For commercial nurseries, sitting a new nursery far away (beyond the insect flight distances) from existing orchid collections, as well as propagating rapidly from seedlings in flasks and from other virus-free plants offer partial solutions.

The main viral diseases of orchids are:

1. *Cymbidium* Mosaic Virus
2. *Cymbidium* Necrotic Ringspot Virus
3. Tobacco Mosaic Virus, 'Orchid' type (O type)
4. *Odontoglossum* Ringspot Virus

The commonest symptoms of viral infection are:

1. Irregular, yellow patches on the leaves (chlorotic mosaic)
2. Pitting, black flecking, streaking or spotting on the leaves (e.g. necrotic mosaic in *Onc. goldiana* var. Golden Shower, *Den.* Pompadour and *Cattleya*)
3. Yellow, brown or black ring patches (ringspot)
4. Colour break in the flowers
5. Bud drop
6. Brown, necrotic streaks on the flowers

A single virus may affect several kinds of orchids, producing different symptoms in each of them and, conversely, different viruses may produce similar symptoms in one host. Multiple infections may be present in one host, triggering a destructive response which would not be seen if only one virus is present in the plant. Thus, Tobacco Mosaic Virus O type may be present in *Cattleya* and *Phalaenopsis* without symptoms, but in the presence of *Cymbidium* Mosaic Virus, brown necrotic streaks develop on the flowers. *Spathoglottis* is particularly

susceptible to viral attacks which produce extensive ringspots and mosaic on the leaves and necrosis of new shoots.

Many viral diseases can be recognised by the symptoms which they cause. To confirm that a plant is infected with virus, the plant tissue can be examined with an electron microscope, or a transmission test can be performed.

The common *Spathoglottis* is very sensitive to virus and a comprehensive, although non-specific, test for the presence of virus involves the transmission of the virus to the *Spathoglottis* plant. To perform the test, a young, virus-free *Spathoglottis* seedling is required, and it can only be used once, after which it must be destroyed. After bruising the upper surface of the test leaf with 3/f grit carborandum applied with a dry watercolour paint brush, spread on it a drop of sap from the orchid plant which is suspected of harbouring virus. Four to six weeks later, if the virus is present, diamond

Top: The magenta speckling on the semi-*alba Cattleya* shows that the plant is infected with virus.

Above: The *Phalaenopsis* has abnormal petals due to bad genes, and the flowers will never improve.

Both plants should be discarded.

shaped spots or rings of necrosis (dead tissue) will appear on the *Spathoglottis*. By using a control plant (i.e. one that is not treated with the sap), a comparison can be made.

In order to identify the type of virus, other indicator plants are used: *Cassia occidentalis*, *Chenopodium amaranticolor* and *Datura stramonium* for *Cymbidium* Mosaic Virus; *Chenopodium amaranticolor* for *Odontoglossum* Ringspot Virus; *Gomphrena globosa* and *Chenopodim quinoa* for Tobacco Mosaic Virus, O type.

Bacterial and Fungal Diseases

For practical purposes, bacterial and fungal diseases can be lumped together because (1) they both appear when there is overcrowding, excessive humidity and heat and (2) the remedy is identical for both conditions. Antibiotics, which play the key role in controlling human bacterial infections, have not been tested on orchids, and at the present time, they are not recommended.

Bacterial and fungal diseases are commonly present as soft, soggy, translucent, circular patches on the leaves. In the case of bacterial rot, the patch enlarges rapidly and soon covers the entire leaf. Gradually, the disease spreads to the roots and pseudobulbs and may destroy the whole plant if prompt treatment is not given. Some fungal diseases are present as yellow spots or black patches on the leaves and stems. These spread quite slowly, and there is ample time to control the infection. In addition to providing good air circulation, protecting seedlings and soft-growing orchids such as *Phalaenopsis* from rain (using discrimination when watering during the rainy season), prophylactic spraying with fungicides is a widely practised method of control. Where infection is already present, the practical steps to take are:

1. Increase the ventilation, if possible.
2. Withhold watering and stop the spraying of organic fertiliser.
3. Isolate badly affected plants.
4. Trim off all badly affected leaves and flowers.
5. If the stem or pseudobulb is affected, prune down drastically to reach clean, healthy tissue and then remove an additional portion of healthy tissue to be on the safe side.
6. Apply a paste of fungicide on the raw edges of leaves, stems and pseudobulbs.
7. Discard any inexpensive, badly affected plant.
8. Discard infected potting medium. If the pots are to be re-used, they should be washed and sterilised by soaking in a 2 percent formalin solution; clay pots may be flamed.
9. Spray the entire collection with at least two fungicides.
10. Finally, check the plant house for leaks and repair them.

Left: Black spots on terete *Vanda* flowers may be due to chemical burn or virus. Without making special tests, it is difficult to exclude a viral problem.

Opposite: Soft rot, in this case crown rot due to *Erwina* or *Phytomonas*, is one of the commonest and most deadly disease of *Phalaenopsis*. It sometimes affects other genera as well. The bacteria is ubiquitous in the orchid's environment, but it is only able to enter the plants when there is a prior injury. Dripping water is the commonest culprit, and water-logging is the commonest predisposing cause.

The entire affected area must be removed, as one would a gangrenous leg, cutting beyond the infected tissues. On this plant, the placement of the cut is indicated by the position of the secateurs.

Bacterial Brown Spot

This is caused by *Phytomonas* (*Pseudomonas*) *cattleya*. It commonly affects compot seedlings, *Phalaenopsis* and *Cattleya*. It appears as a circular, water-soaked spot which enlarges rapidly and is light brown in the centre with a rim of dirty green. The disease spreads so swiftly in *Phalaenopsis* that within one or two days it may reach the stem and kill the plant. The infected tissues are soft and friable, breaking up to release a watery exudate which is a pure culture of bacteria. It is probably spread through insect bites or persistent drip on the foliage.

Diseased plants must be carefully lifted and isolated. Affected parts are cut off, along with a generous zone of the surrounding healthy tissue. Soak the entire compot or plant in Natriphene, 1:2,000 solution for 5–6 hours, allow to dry and spray with residual fungicide such as Tersan. Isolate the treated plant for a few weeks until certain that no new patch of bacterial brown spot appears; it is then safe to return the plant to its old position. An alternative method is to use Physan, 0.5–2 teaspoons per 4 litres of water, sprayed generously. Repeat a week later.

Bacterial Soft Rot

Like bacterial brown spot, this is a rapidly spreading disease which presents itself as a water-soaked, elevated spot on the top surface of the leaves, usually of *Phalaenopsis*. The surface appears wrinkled, but the tissues are friable underneath, and on breaking up, it releases the bacillus, *Erwinia carotovora*, which produces a foul smell. There is no cure, and the plant must be destroyed immediately. Sterilise the surrounding area by pouring over with methylated spirit and flaming, and treat the neighbouring plants with fungicide.

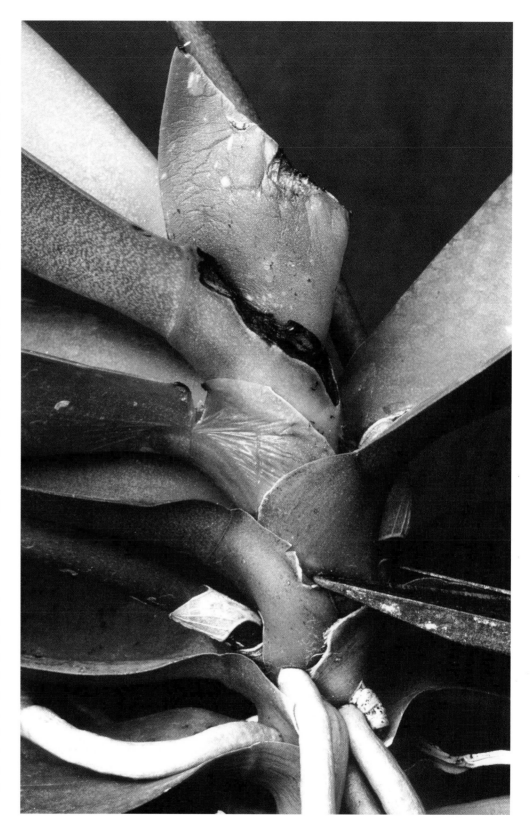

Bacterial Leaf Rot

This is a condition related to the above. *Erwinia chrysanthemi* has been found to be the cause of leaf rot among orchids in Hawaii. In *Dendrobium*, the leaves become yellow and water-soaked, while in *Vanda*, it shows as translucent patches on leaves which become darkened, black and sunken. The treatment is the same as for bacterial brown spot.

Brown Rot of *Paphiopedilum*

Another species of *Erwinia*, *E. cypripedii*, is responsible for this destructive disease which can kill a plant without prior warning. The first hint is a chestnut brown spot on the leaves. It is introduced through a cut or insect damage and spread rapidly in both directions, entering the crown and spreading through the rhizomes. If recognised early, the rhizome should be severed, the affected growth removed and discarded, the healthy portion soaked in fungicide and then repotted. Maintain a strict fungicide programme thereafter.

Crown Rot, Black Rot, Damping Off

Among the fungal diseases, this is the most serious. It is caused by *Phytophora omnivora*, the fungus responsible for the tragic potato blight that afflicted Ireland in 1845, and *Phytophora palmivora*, which is prevalent in Southeast Asia and is the cause of crown rot in Singapore. The disease appears during the rainy season and commonly affects semi-terete *Vanda* and *Aranda* which are grown in the open and *Phalaenopsis* growing in lath houses. The bad practice of spraying the entire orchid including the crown with organic fertiliser is another contributory factor.

Phytophora palmivora does not always produce crown rot when it is present. Initially, the infection may appear on the young leaves as small, watery brown spots surrounded by a yellow margin. If the affected parts are removed and the plant is sprayed with fungicide, crown rot can be prevented. During prolonged wet weather, the infection spreads rapidly into the crown, causing the leaves to turn black and rot. These leaves are easily lifted off the plant. The infection extends downwards rapidly, piercing the stem and causing the plant to die. The disease spreads rapidly to the adjacent plants through splashing water.

Phytophora palmivora and several species of *Pythium* are the main causes of 'damping off' of compot seedlings in Singapore. Severely affected plants are best destroyed.

Root Rot

This is caused by *Pellicularia filamentosa*, a brown fungus which infects the roots, rhizomes and lower portions of the pseudobulbs. It can destroy seedlings in a very short time and cause adult plants to wilt, turn yellow and die

Left: *Phalaenopsis* leaf affected by *Anthracnose*

Right: Floral Blight on *Dendrobium* caused by fungus

Overcrowding and poor wind movement are predisposing factors.

for no obvious reason. Affected plants should be soaked in 1:2,000 Natriphene solution, or Thiram (25 g per 10 litres water) or Zineb (25 g of 80 percent wettable powder per 10 litres water). The treatment should be repeated every 4–7 days.

Anthracnose or Leaf Spot Fungus

This is a common but usually non-lethal fungal disease which causes damage to flowers and leaves of a wide range of orchid plants throughout the world. It is caused by various species of *Colletotrichum* and *Gloeosporium*. The leaves develop brown, transverse, concentric rings or bands across their distal portion, and flowers are spotted with small, dark brown and raised black pustules on the posterior surface of the petals and sepals which then become twisted and deformed.

Infected flowers and leaves should be removed. The plant should then be sprayed with Maneb, Zineb or Ferbam (25 g 80 percent wettable powder per 10 litres water).

Fusarium Flower Blight and Leaf Spot

The causative organisms, *Fusarium moniliforme* and *F. oxysporum*, attack the plant through the roots and cut ends of the stem. In Singapore, *F. moniliforme* is one of the major causes of bud drop, but only in its more serious forms, leaf spot or *Fusarium*, does it result in wilting. Toxic substances produced by the fungus cause a plugging of the vascular bundles of the leaves and stem, denying water to tissues, and this causes the flowers and leaves to turn yellow and wilt. Surface lesions become sunken, dark brown or black, and they are often covered with powdery, white mycelia or numerous, tiny, pinkish spore masses. When the rhizomes and pseudobulbs are affected, a pink-purplish colour may travel through the infected tissues.

Infected inflorescences must be removed and the plant sprayed with Benomyl (15 g 50 percent wettable powder per 10 litres water) every 4–7 days.

Curvularia Flower Blight

This appears as numerous, tiny, dark brown spots on the flowers. Similar flower speckling may also be caused by *Botrytis cinerea* and *Sclerotina fucheliana*. Treatment is by spraying with Zineb, Ferbam or Maneb. The flowers must not be sprayed with Captan, which may injure the flowers.

Cercospora Leaf Spot

Several species of *Cercospora* cause leaf spots in a variety of orchids, but in Singapore it appears to be confined to *Dendrobium*. Circular yellow spots first develop on the undersurface, but they are usually not detected until they have spread to the upper surface. Eventually, the spots become dark and sunken and coated with a black, powdery film of spores on the under surface of the leaves.

Remove all the affected leaves and spray with fungicide. Use fungicides in rotation. The common fungicides are Zineb, Maneb, Ferban, Captan, Thiram, Santar and Consan.

Phyllostictina Leaf Spot

This is caused by *Phyllostictina citracarpa*. At a casual glance, it resembles the mosaic pattern common to viral diseases. However, careful examination will reveal the dark spots which are the fruiting bodies of the fungus that runs throughout the affected area. If one is uncertain of the cause of the lesion, wipe the leaf with a damp cloth; fungal spores will come off as a dark, fine powder.

The treatment is the same as for *Anthracnose*, outlined above.

Snow Mould

Many old plants and neglected hanging baskets of *Vanda* are often covered with a white powdery growth that completely covers all the old roots. This is due to *Ptychogaster spp*. The fungal covering is impermeable to water, and as it completely covers the roots right to their tips, they become suffocated and dehydrated. None of the nutrients sprayed on such roots ever reaches the plant.

Repot the plant after soaking in Shield 10, 2 ml of concentrate per litre water, or a similar fungicide. Spray regularly with Shield 10 to prevent reappearance of the fungus. Discard old pots and baskets which are covered by the fungus.

Sooty Mould

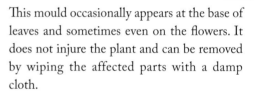

This mould occasionally appears at the base of leaves and sometimes even on the flowers. It does not injure the plant and can be removed by wiping the affected parts with a damp cloth.

Pests

In the ecological balance of nature, mites, insects, slugs and snails prey upon orchids at least to the same extent that they do upon other plants. They have been distributed worldwide through the import of infested plants long before quarantine laws came into effect. In the open cultivation practised in the tropics, orchids are subjected not only to host-specific pests but also to incidental chance depredation by other parasites which normally infest fruit trees, ornamental shrubs and other plants. Thus, unlike their fortunate counterparts in the temperate countries of Europe, North America and Australia who need not apply insecticides and miticides more often than twice a year (because their greenhouses are completely free of pests), the tropical orchid grower has to maintain a strict rotation of chemical sprays, particularly if he is growing for cut flowers.

The extent and type of damage caused by an orchid pest depends on the severity of the infestation and on the feeding habits of the particular pest. Mites and insects generally cause damage to foliage, flowers and stems, while slugs and snails destroy roots and foliage. Burrowing and sucking insects introduce serious viral, bacterial and fungal diseases, and they are much more serious than caterpillars or grasshoppers, which merely chew off leaves and flowers. Their control in the tropics is based on (1) an understanding of their life history and (2) whether the grower intends to supply the florist with his flowers.

To meet the high standards required by the florist, the flowers, at least, if not the entire nursery, must be free of insects and mites. Spraying with insecticides and miticides must be done frequently enough to destroy all larvae emerging from deposited eggs before they reach the reproductive stage; this prevents a build-up of any infestation. The eggs of most mites and insects hatch in 3–5 days, so between the first and second spraying, there should be a strict interval of five days. This will eradicate existing infestation and the potential for future build-up. However, unless the pesticides are systemic, they will not protect the plants against newly laid eggs deposited, whether between the two sprayings or subsequent to the second spraying. Since it takes a minimum period of one week from hatching for a mite or insect to mature, a third spraying should be made 10 days after the first spraying (3+7), and the schedule repeated. This is too complicated to fit into a weekly schedule, and in practice most commercial nurseries in Singapore will spray pesticides twice a week, with some nurseries resorting to three applications a week.

Amateur growers should not resort to frequent prophylactic spraying of pesticides and should work out individual programmes to suit their own growing environment, perhaps spraying twice (five days apart) every 3–4 weeks.

Spider Mites (*Tetranychidae*)

Mites are eight-legged arthropods, and both the spider mites (red spiders) and the false spider mites are serious pests on all tropical orchids. The spider mite spins fine webs which may be detected on the flower spikes and along the edges of leaves and floral segments. They are about 0.5 mm long, just visible to the naked eye as moving red spots when the infested plant is examined closely; they are better detected by means of a hand lens. They hide from the sun, feeding on the undersurface of the leaf, causing it to become speckled.

The female lays one or two eggs a day, and these hatch in 3–5 days. The mite becomes mature in a week.

The best control is Dimite (diparachlorophenyl methylcarbinol or DMC), a compound closely related to DDT. Malathion and Roxion are also effective.

False Spider Mites

False spider mites cause severe destruction of affected foliage with pitting, chlorosis and death of the leaves. They are particularly disastrous for *Phalaenopsis*. They are about half the size of red spider mites and are better seen

Left: *Phalaenopsis* leaf infested with spider mites whose colonies show up as tiny red spots

Opposite: The original (smaller) plant was attacked by a weevil which laid its eggs at the growing tip of the plant and killed off its apical meristem. After the initial setback, a new (larger) plant has grown up on the side. With crown rot, the entire plant would have been lost.

through a hand lens. The female lays one or two bright red eggs which hatch in a week, and the mites take two weeks to mature.

The best control is with Dimite, Kelthane or Chlorobenzilate. They are not killed by Malathion. Hot, dry weather encourages their development, and frequent fine spraying with water helps to keep their population down.

Thrips (*Thysanoptera*)

Thrips are small flying or jumping insects about 1 mm long. They feed on leaves and flowers, causing fine silver specks which completely ruin the flowers. They commonly attack flower buds of vandaceous orchids and hide within them, in the meantime sucking the juices from the unopened sepals and petals which are disfigured when they open. The insects are yellow-grey or black, and their presence is easily recognised by the black excrement which they deposit on the infested parts.

The female lays 25 eggs which hatch in 2–3 weeks. The thrips reach maturity after three weeks. Several species attack orchid plants.

Many insecticides, including Roxion, DDT and BHC (benzene hexachloride, lindane), will easily eliminate thrips, but perhaps the cheapest and safest is Dipterex (trichlorophon) which is biodegradable.

Scale Insects

Soft scales (*Coccidae*) and armoured scales (*Diaspididae*) commonly affect orchids, but they generally do little damage compared with mites, thrips and weevils. They are found on the undersurface of leaves and on the stem. The young stages are free-moving, but when they mature they stick their proboscis into the tissues and remain there. The soft scales have a flattened, elliptical, waxy covering and are about 0.5–1 cm in diameter. The soft scales produce living young while the armoured scales emerge under the protection of the mother's shell. The scale insects take a long time to develop, and only a few generations are produced a year.

Good control is provided by spraying with Malathion or white oil (Albolinum). Repeated treatment over several weeks may be necessary to destroy all the young which are continually produced by the slow-hatching eggs.

Aphis (*Aphidae*)

Aphids or plant lice are soft-bodied green or black insects which are incidental pests. They suck the sap and excrete honeydew on new spikes, flowers and foliage. Their principal threat to orchids is serious because it is insidious — they transmit viral disease. The female produces living young which take a week to mature. Malathion provides satisfactory control.

Mealy Bugs (*Pseudococcidae*)

These are incidental pests sometimes affecting *Phalaenopsis*. Although they are common on certain fruit trees (such as Annona), they are not common on orchids in Singapore.

The adults are 0.5–1 cm long and are covered with a whitish powder and long, waxy filaments. The eggs are produced in cottony egg sacs and hatch in 4–10 weeks, becoming adults in 2–3 months. Parathion provides the best control but Malathion is satisfactory.

Weevils

These are a great scourge because the grubs bore through the growing stem and destroy the entire plant. The adult weevils are 0.5–1 cm long, black and easily recognised by their long snouts. They are best detected in the early morning when they come out to feed. Several generations are produced a year.

Beetles

A yellow orchid beetle is sometimes seen in Singapore preying on orchids. It lays its eggs on the flowers, and the slimy, brick-red grubs feed on the flowers. They grow to a length of 1.5 cm and spin a thick white cocoon along the rachis and on the stems. The damage is usually restricted to the flowers; cutting off entire sprays and burning them appears to be a simple way for control. If the plants are regularly sprayed with pesticides, these beetles are seldom seen. Long-horned beetles are also reported on orchids in Southeast Asia, and bug beetles attack *Dendrobium* in Papua New Guinea.

Caterpillars

A good pesticide programme generally keeps the butterflies at bay. A variety of butterflies lay eggs on *Phalaenopsis* seedlings and on the developing buds of other vandaceous orchids and *Dendrobium*. The caterpillars bore into the buds and cause extensive damage to the flowers. Once infested, the entire spray should be removed and burnt.

Snails

Garden snails attack orchids within their reach. The tiny bush snails which hide in the charcoal and crock within the pots are particularly troublesome. They eat up young leaves, flowers and the tender root tips. Methaldehyde granules applied around the plant weekly over three weeks will eliminate the menace.

Another method for control is to put out deep wide-mouth containers holding stale beer in the evening. The snails are attracted to the beer and drown in the liquid.

Pesticides

A large variety of pesticides have been developed to control agricultural pests and disease-bearing insects, but only a few of these have been tested on orchids. Their suitability for orchids is always printed on the commercial packing which also gives the instructions for their use. These instructions should be followed. For use on orchids, pesticides must be dissolved only in water; the solvents used for household insecticides, for example, are toxic to plants.

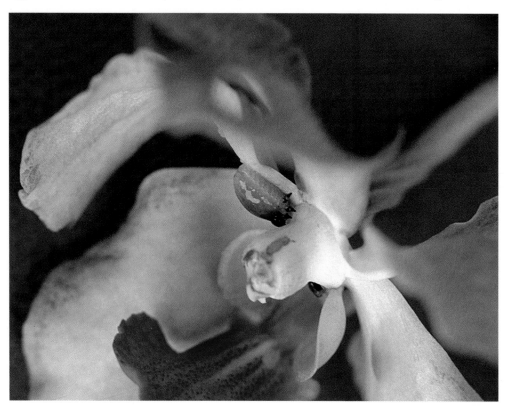

All insecticides are poisonous to man and must be handled with care. Some are so poisonous that they must never be used by the amateur or even by commercial growers inexperienced in their handling. On occasion, some agricultural organisations may arrange for the spraying of a potent pesticide if a particularly serious infestation necessitates its use; in such an event, highly-trained personnel wearing special protective clothing and gas masks are employed. With regards to the insecticides which are available across the counter, the degree of safety associated with their usage depends on two main factors: (1) the lethal dose, with LD50 as a reference number (this means a 50 percent kill on test animals expressed as mg pesticide per 1 kg body weight of test animal) and (2) whether the pesticide is biodegradable or whether it has a cumulative effect. The higher the LD50,

the safer the pesticide, because it requires a larger dose of it to kill the test animal. Those at 5,000 are relatively safe. Those at 500 are also safe, but should be handled with caution while those between 50 and 500 are quite poisonous and should only be used if nothing else which is safer will work.

Organic phosphates are nerve poisons which are extremely poisonous to man, with the exception of Malathion. Because they are all cumulative poisons, even Malathion must be handled with care. Benzene hexachloride

(BHC), DDT and Dimite (DMC) are chlorinated hydrocarbons that have a cumulative effect, DDT being particularly harmful on the environment. Their toxicity varies from being relatively harmless (Aramite) to moderately toxic (benzene hexachloride). DDT is banned in many countries but is still used in some countries where the control of malaria is a prime concern. Dipterex is relatively safe. Plant derivatives such as rotenone (Derris), and now Nim Oil, are relatively non-toxic.

Opposite: *Vanda* flower attacked by the yellow beetle. The beetle had laid its eggs on the flower when it first opened, leaving the larvae to feed on the flower. The picture shows a beetle ready to leave its cocoon.

Right: Sunburn of *Phalaenopsis*

Bottom, Left: Chemical burns

Bottom, Right: Sunburn

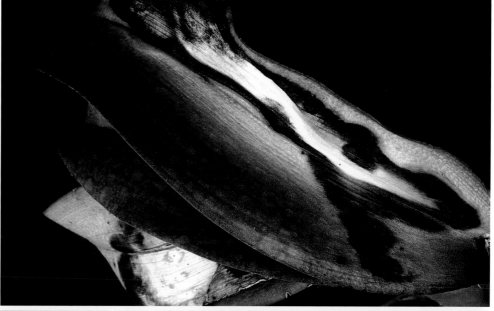

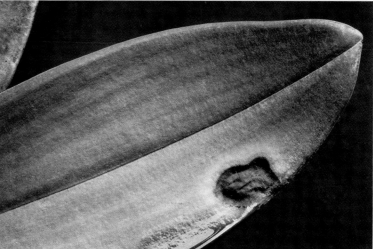

Chapter 14

Handling Cut Flowers

Anyone who has ever placed fresh flowers in a vase must have pondered on how best to extend their vase life. The common practices are:

1. Cut in the morning or evening, not in the noonday sun.
2. Make a fresh cut before placing the flowers in the vase.
3. Select half-open blooms or sprays which still have some unopened buds.
4. Use clean containers.
5. Add aspirin to the water.

We also know that flowers last longer in temperate climes than in the tropics.

Because of the importance to the cut-flower industry, considerable research has gone into the behaviour of flowers after their harvest. These studies show that the vase life of orchids can be shortened or prolonged by:

1. Growing conditions
2. The harvesting process
3. Handling of the orchids after the harvest

Before discussing these factors, we should perhaps point out the obvious. that vase life is predetermined mainly by the type of orchid and, within a cross, by the actual clone. For

Left: A new white curly-petal *Dendrobium* that would do very well as a cut flower if the inflorescence carries many flowers

example, while *Aranda* may last four weeks, in the same atmosphere, *Cattleya* will last only four days. Among the *Aranda*, Christine No. 1 has a shelf life of four weeks, but No. 999 lasts only five days.

Growing Conditions

Weak plants produce poor-quality flowers which do not last. Never buy poor quality flowers, even if they are cheap. If you do not wish to spend much, buy one spray. But it must be of top quality.

It is estimated that a third of the post-harvest life of flowers is preprogrammed by preharvest handling. Optimum growing conditions ensure that the stems are stout and capable of transporting large volumes of water to the flowers. The flowers themselves will have heavier substance and are free from pests and disease. Good nurseries have a preventive spraying programme which ensures that the plants and flowers are always free of disease.

Harvesting

First, let us consider the selection of flowers. If one is choosing a single bloom, for instance a *Phalaenopsis* or *Cattleya*, it is necessary to be sure that the bloom is at least three or four days old. Before the flower has fully opened, its tissues are weak, and a flower removed prematurely will not reach its full size, is prone to wilting and will not last. At the other extreme, one would also not select a bloom which is already showing evidence of senescence (ageing).

When choosing an orchid spray, the best time to harvest is when the spray is 60–70 percent bloomed. At this stage, it is in top condition: sufficient flowers have opened for one to appreciate their arrangement on the spray, and premature colour fading has not occurred in the first blooms. Furthermore, the rest of the spray will open over the next few weeks if the flowers are properly handled.

A spray that is fully opened is never quite as attractive because its older flowers are already half spent, and it will not achieve maximum vase life.

On the other hand, a spray which is only one-third open or still in bud is unlikely to bloom to the last flower, even if it is expertly handled. Unlike roses or lilies, orchids cannot be harvested in bud.

The spray should be cut with a sharp blade, as close to the stem as possible. Tissue injury at the base of a crushed stem promotes enzyme reaction and bacterial rot which clog up the water channels. A long stem is essential to facilitate post-harvest handling which will involve repeated freshening of the cut end, i.e. making a fresh cut every two or three days. In fact, orchid breeders who have an eye on the cut-flower market are constantly breeding for long flower stalks.

During harvesting, 'field heat' is an important consideration. The temperature of the flowers is higher in the early afternoon; the stomata are open to promote transpiration and cooling; the flowers are not so turgid. If the flowers are cut and not immediately placed in water, they tend to wilt more easily. More air will enter the water channels and cause blockage. Heat also promotes the growth of micro-organisms. Field heat can be removed by placing the blooms in a refrigerator for 20 minutes.

Dark flowers are particularly stressed. They may be as much as 5.5 degrees Centigrade warmer than white flowers during the mid-afternoon. If one is running a nursery, the darker flowers should be harvested in the morning, the white flowers later in the day.

Overheating can also occur if flowers are bunched together and too quickly packed into a box. Flowers should preferably be allowed to cool down or be stored at 10–15 degrees Centigrade before packing.

Post Harvesting

On a plant, the flowers of *Vanda* Miss Joaquim fade in turn. If the entire spray is placed in a box, however, they fade simultaneously. When several sprays are placed together into a box and there is just one damaged flower, all the flowers may have faded when the box is opened the following day. If one packs an apple into a box of *Vanda* Miss Joaquim, the flowers also fade overnight.

The cause of simultaneous fading of the flowers is a gas, ethylene dioxide, which is released during the ageing of orchid flowers. It does not affect all orchid flowers the same way; some flowers, such as *Dendrobium* Pompadour, may, in fact, have their vase life extended when they are exposed to small amounts of ethylene.

The release of ethylene may also be triggered by the removal of pollen. Sometimes, even the mere removal of the pollen cap will do it. *Vanda* and *Cymbidium* are particularly sensitive to ethylene.

In the handling of orchids, one should ensure that the pollen is not accidentally removed. If a spray is to be packed into a box, it is important to spot those flowers which have lost pollen and to exclude them from the box. Our advice is to examine your flower arrangement every morning and to remove those flowers which have faded, or they may spoil the others.

The other factors which influence the shelf life of orchid cut flowers are:

1. Water balance
2. Nutrition
3. Bacterial rot

Small tubes run along the stem towards the flowers, conveying water to them in much the same way that blood vessels convey blood to distant tissues in the human body. Obviously, the stouter the stem, the more tubes there are, and the better the spray will withstand wilting. Blockage of the tubes at the entry point or higher up is regarded as a major cause of wilting. Many florists advise cutting rose stems under water to prevent air bubbles from clogging up the stems. This is not necessary with orchids, which can be cut and then placed into water. If the sprays have been left out of water for several hours, a fresh cut should always be made before placing the flowers in water.

Micro-organisms like bacteria and fungi block the stem by growing on it or, indirectly, by producing enzymes which cause degeneration of the cell walls leading to the formation of plugs. Low pH (caused by adding acid or aspirin to the water) or the addition of preservatives such as sodium azide or hydroxyquinolene prevents bacterial growth and blocks enzyme activity, both of which may extend the vase life of flowers. Spraying flowers with antiseptic solutions (hexa-chlorophene) is known to prolong their vase life. The spraying of organic fertilisers just prior to harvesting leaves a residue on the flowers and may encourage the growth of micro-organisms.

Plunging the orchids for 10 minutes into 100–1,200 parts per million of silver nitrate may prolong the vase life of cut orchids by 20 percent. Jane Goh and Dr. Goh Chong Jin found that it prolonged the vase life of *Arachnis* Maggie Oei in Singapore by three days.

Water quality may be important. Clean tap water, fit for drinking, contains many dissolved salts which may be harmful to flowers. Fluoride, for example, though excellent for teeth, is not good for flowers.

It has been shown that stem-plugging decreases with water purification. The various methods of water purification are:

1. Distillation in a copper still
2. Glass distillation
3. Passage through a deioniser
4. Passage through a micropore filter

Of these water treatment methods, perhaps only deionisation is practical for private or commercial purposes.

The other component in water balance is water loss. Low humidity within an air-conditioned room encourages water loss, whereas storage at low temperature and transport in sealed plastic bags prevent water loss. If water stress is present and the flowers are limp on arrival, the entire inflorescence should be immersed in lukewarm water for an hour or two or until the flowers regain their turgidity.

Sugar solutions induce closure of stomata in cut roses, resulting in water retention and gain in fresh weight, even though the roses actually absorb less solution. It is not the case with snapdragons. Sugar also has a nutritional role and, by slowing the breakdown of protein and its associated ammonia release, it can extend the vase life and attractiveness of roses. Whether the same applies to orchid flowers has not been demonstrated.

It is well known that flowers keep better at lower temperatures. But with tropical orchids, chilling injury occurs if the flowers are stored at 4 degrees Centigrade for extended periods. The flowers, indeed entire plants, turn to mush if kept at 4 degrees for too long. The best temperature to store tropical orchids is at 10 degrees Centigrade. Low temperature slows down the respiration rate, growth of organisms and enzymatic reactions.

For orchid exporters, the practice of controlled or modified atmosphere storage is worth looking into. This aims at reducing water loss, reducing respiration to conserve food reserves and delaying the effects of ethylene. At the University of Malaya, Helen Nair conducted experiments on *Dendrobium* Pompadour. She found that storage under 10 percent carbon dioxide for four days extended the bench life by four days.

Removal of oxygen is also beneficial because it slows down respiration. For packing purposes, it is easy to substitute air with a mixture of nitrogen and 10 percent carbon dioxide. The actual removal of ethylene involves the use of absorbents such as activated charcoal or potassium permanganate, and/or storage in low pressure compartments.

Below: Cut sprays of *Aranda* Christine No. 1 being seasoned in buckets of water treated with preservatives. What happens at this stage and when the flowers were harvested are possibly the most important factors that will determine the vase life of the flowers when they are presented to you at the flower shop.

Glossary

Aerides	A popular Asian genus of monopodial orchids
aggregates	Clay particles used in hydroponic culture
allotetraploids	Amphidiploids
amphidiploids	Hybrids of two species or genera possessing all the chromosomes of the parents or twice the number of chromosomes. Such plants are important building blocks in hybridisation.
anak	A Malay word for 'baby', referring to plantlets
aneuploid	Having more or less than an integral multiple of a haploid number of chromosomes (for example, $2n \pm y$, $3n \pm y$, $4n \pm y$, etc.)
anther	The part of the flower bearing the pollen; the top end of the column
apical	At the apex of the pseudobulb
Arachnis	A genus of Scorpion Orchids
Aranda	An artificial hybrid genus produced by crossing *Arachnis* and *Vanda*
Asocenda	'Miniature *Vanda*' produced by crossing *Ascocentrum* with *Vanda*
asymbiotic	Referring to the laboratory culture of orchids from seed which has bypassed the need for a symbiotic relationship with fungus
axillary	At the axil of the leaf
back bulb	An old live pseudobulb, usually leafless and shrivelled, at the original end of the rhizome
basal	Arising at the base of the pseudobulb, usually referring to the inflorescence
bi-generic	Hybrids between two genera
bract	Leaf-like structure at the base of the flower or side branch of an inflorescence
cane *Dendrobium*	Antelope or horn *Dendrobium*, members of the section *Ceratobium*
chromosome	A rod-like structure composed of DNA which carries the genetic code; it is located in the nucleus and becomes visible at mitosis
clone	Exact duplicates of a plant produced vegetatively, such as by cutting; nowadays commonly produced by meristemming
column	The organ located at the centre of the flower and produced by the infusion of male and female reproductive parts.
community pot	A simple pot growing seedlings, newly planted out from flasks, together
compost	Leaves or cut grass used for covering the feeding roots of monopodial orchids which are planted on the ground
crock	Broken pots, brick or charcoal placed at the base of a pot to facilitate draining
Cymbidium	An important genus of Asian orchids widely cultivated in the temperate region and most popular with the Chinese
damping off	Soft bacterial rot affecting seedlings
deciduous	The habit of shedding leaves at the end of the growing season
Dendrobium	Popular Asiatic genus of sympodial orchids comprising some 600 species divided into 20 sections
diploid	Possessing two complete sets of chromosomes
diurnal	Day and night rhythm
Doritis	Asian genus closely related to *Phalaenopsis* and freely interbreeding with the latter; it has only one species.
dorsal	Back or uppermost
embryo	Miniature plant contained in the seed
epiphyte	A plant which grows on another plant without deriving its nourishment from the latter
equitant	Describing leaves which overlap at their base and form two neat rows like an open fan, as in *Oncidium*.
Euanthe	A proposed botanical genus containing the single species, *Vanda sanderiana*
fertilisation	The fusion of the pollen nucleus with the ovule nucleus; it occurs several days to several months after pollination, depending on the species of orchid
foot	Projection at the base of the column, often extending to the lip

gamete	The pollen cell or ovule
genus	A taxonomic grouping of species, all sharing certain common characteristics.
habit	The general appearance of the plant, whether erect, climbing, pendent, etc.
hirsute	Hairy
inflorescence	Flower spike
keiki	A Hawaiian word for 'baby', like the Malay term *anak* used by orchidists to refer to plantlets
labellum	Lip
lip	A modified petal in the orchid, highly distinctive and colourful
lithophytic	Growing on rocks
mentum	Chin, formed by the union of the column foot with the basal, inner margins of the lateral sepals
meristem	Undifferentiated tissue found at apices of stems, buds, roots and young leaves
mitosis	Cell division.
monopodial	Referring to plants that continue to grow from the apex
mycorrhiza	The root fungus, a universal symbiont of the orchid; it penetrates the seed coat and provides sugar to the germinating embryo
Oncidium	A large Central American genus known as the Dancing Ladies
osmunda	Fibrous roots of osmunda fern, used as a potting medium for seedlings
ovule	The female gametes in the seed pod which become the seeds when fertilised
pannicle	An inflorescence with side branches, as seen in *Renanthera*
Paphiopedilum	The Asian Slipper Orchid
petal	The inner whorl of the perianth; in orchids, there are only two petals, the third being modified to form the lip
Phalaenopsis	The beautiful Moth Orchids of Southeast Asia
polyploid	Having three or more full sets of chromosomes
pseudobulb	The succulent stem of a sympodial orchid
raceme	An unbranched inflorescence, as seen in *Rhynchostylis* (cf. pinnacle)

Renanthera	A delightful Asian genus of monopodial orchids which bear large pinnacles of red flowers
rhizome	A horizontal stem which bears erects branches (pseudobulbs) and roots
Rhynchostylis	The Foxtail Orchids native to Thailand
rostellum	A 'beak'; the tip of the column, it separates the stigma from the pollinia above
saprophyte	A plant which delivers its sustenance from decomposed organic material
semi-terete	Referring to the thickened, grooved leaves of hybrid *Vanda* which have been produced by crossing terete with strap-leaf *Vanda*
sepal	The outermost whorl of the flower. Orchids have three sepals, all having the same colour as the petals.
Spathoglottis	A ground orchid, extremely widespread in Asia and one of the first colonisers of the land
species	A distinct grouping of plants within a genus, all members being interfertile with one another
spur	The tubular extension of the lip which contains the nectary, as in *Dendrobium*
staminode	A sterile stamen, without pollinia
stigma	The part of the column which receives the pollen
strap-leaf	Leaves which are long and flat, resembling a leather strap
substance	Referring to the thickness of the petals and sepals, usually an indication of their longevity
sympodial	'Feet together', an orchid which has a horizontal stem or rhizome which produces lateral branches or pseudobulbs that bear flowers and leaves
tepals	Referring to petals and sepals which look alike
terete	Pencil-like, describing leaves
terrestrial	Growing in the ground
tribe	A large grouping of genera which have certain similarities
Vanda	A great monopodial genus from Asia
variety	A distinct plant of some horticultural merit which may be designated by a varietal name
velamen	The thick layer of dead cells which cover the roots of epiphyte orchids

Selected Bibliography

Assavapiches, C. *Beautiful Orchids*. Bangkok: C.C. Chao Orchid Garden, *circa* 1977.

Black, P.M. *Beautiful Orchids*. London: Hamlyn Press, 1973.

Blowers, J.W. *Pictorial Orchid Growing*. Self-published. 96 Marion Crescent, Maidstone, Kent, England, 1966.

Cady, L. and E.R. Rotherham. *Australian Native Orchids in Colour*. Rutland: Charles E. Tuttle, 1970.

Cootes, J. *The Orchids of the Philippines*. Singapore: Times Editions, 2001.

Cribb, P. *The Genus Paphiopedilum*. London: Royal Botanic Gardens, Kew, and Collingridge Books, 1997.

Davis, R.S. and M.L. Steiner. *Philippine Orchids*. New York: William Frederick Press, 1952.

Hawkes, A.D. *Encyclopaedia of Cultivated Orchids*. London: Faber and Faber, 1965.

Henderson, M.R. and G. Addison. *Malayan Orchid Hybrids*. Singapore: Government Printing Office, 1957.

Holttum, R.E. *Flora of Malaya, Volume 1 — Orchids*, 3rd ed. Singapore: Government Printing Office, 1964.

Kamemoto, H. and R. Sagarik. *Beautiful Thai Orchid Species*. Bangkok: Orchid Society of Thailand, 1975.

Latif, S.M. *Bunga Anggrik. Permata Belantara Indonesia* (in Bahasa Indonesia). Bandung: N.V. Penerbitan W. van Hoeve, 1953.

Noble, M. *You Can Grown Phalaenopsis Orchids*. Self-published. 3033 Riverside Avenue, Jacksonville, Fla. 32205, USA, 1972.

Northen, R.T. *Home Orchid Growing*, 3rd ed. New York: Van Norstrand Reinhold, 1970

O'Byrne, P. *A to Z South East Asian Orchids Species*. Singapore: Orchid Society of Southeast Asia, 2001.

Richter, W. *Orchid Care*. London: Van Norstrand Reinhold, 1972.

Ridley, H.N. *Malayan Orchids*. Singapore: Botanic Gardens, *circa* 1900.

Royal Horticultural Society. *Sander's List of Orchid Hybrids*, Addendum I and II. London.

Sanderson, F.R. and T.A. Yong. *Diseases of Orchids in Singapore*. Singapore: Primary Production Department, Ministry of Law and National Development, 1972.

Seidenfaden, G. and T. Smitinand. *The Orchids of Thailand*, Volumes 1–4. Bangkok: The Siam Society, 1959.

Swinson, A. *Frederick Sander, the Orchid King*. London: Hodder and Stoughton, 1970.

Teoh, E.S. (ed.). *Orchids*. Singapore: The Orchid Society of Southeast Asia, P.O. Box 2363, Singapore, 1978.

The New Phalaenopsis of Taiwan II. Taiwan R.O.C: Sogo Orchids Nursery, 1997.

The New Phalaenopsis of Taiwan III. Taiwan R.O.C: Sogo Team Co. Ltd Nursery, 2003.

Vaddhanaphuti, N. *A Field Guide to the Wild Orchids of Thailand*. Chiang Mai: Silkworm Books, 2001.

Wareing, P.F. and I.D.J. Philips. *The Control of Growth and Differentiation in Plants*. Oxford: Permagon Press, 1970.

Waters, V. H. and C.C. Waters. *A Survey of the Slipper Orchids*. Shelby, N.C.: Carolina Press, 1973.

Williams, B.S. *The Orchid Grower's Manual*. 1894. Reprint. New York: Wheldon and Wesley Ltd, 1961.

Withner, C.L. (ed.). *The Orchids. A Scientific Survey*. New York: Ronald Press, 1959.

Withner, C.Ll (ed.). *The Orchids. Scientific Studies*. New York: John Wiley and Sons, 1974.

Proceedings of the Fourth World Orchid Conference, 1963. Burkhill, H.M., B.C. Yeoh and R. Scott (eds.). Singapore: The Straits Times Press.

Proceedings of the Fifth World Orchid Conference, 1966. De Garmo, L. R. (ed.). Long Beach: available from the American Orchid Society Inc.

Proceedings of the Sixth World Orchid Conference, 1969. Corrigan, M.J.G. (ed.). Sydney: available from the American Orchid Society Inc.

Anales de la 7 Conferencia Mundial de Orquideologia, 1972 (in Spanish and English). De Ospina, H.B. (ed.). Medellin: available from the American Orchid Society Inc.

Tagungsbericht de 8 Welt Orchideen Konferenz, 1975 (in German and English). Senghas, K. (ed.). Frankfurt: available from the American Orchid Society Inc.

Periodicals:

American Orchid Society Bulletin

Malayan Orchid Review

NaPua Okika O Hawaii Nei

National Geographic Society Magazine, vol. 139(4): pp. 485–513, 1971.

Orchid Digest

Orchid Review

Index